D1534927

Whose Holy Land?

Archaeology Meets Geopolitics in Today's Middle East

Whose Holy Land?

Archaeology Meets Geopolitics in Today's Middle East

Kenneth L. Hanson

Published by New English Review Press
a subsidiary of World Encounter Institute
PO Box 158397
Nashville, Tennessee 37215
&
27 Old Gloucester Street
London, England, WC1N 3AX

Cover Art & Design by Kendra Mallock

ISBN: 978-1-943003-40-2

First Edition

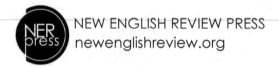

NEW ENGLISH REVIEW PRESS
newenglishreview.org

To Elena, my endless source of inspiration.

The air over Jerusalem is saturated with prayers
and dreams
like the air over industrial cities.
It's hard to breathe.
And from time to time a new shipment of history arrives
and the houses and towers are its packing materials.
Later these are discarded and piled up in dumps.
And sometimes candles arrive instead of people
and then it's quiet.

—*Yehuda Amichai*

Contents

INTRODUCTION

"Only the Jews are expected to be the only real Christians in the world." —Eric Hoffer, 1968.

S O, THE JEWS TOOK Arab land? Are we sure about that? Or was it the other way around? Unfortunately, it is not particularly easy even to ask such questions in today's world. Worse still, there has for some time been a "new anti-Semitism" rising across the globe, especially prevalent in Europe and the Middle East, that will not even allow such questions to be raised. It is painfully evident that the Jewish people have been demonized down through history, something to which they are well accustomed. Today, however, there is a new dimension to this demonization, given that the brutal characterization of the state of Israel prevalent in the Islamic world has now become the dominant European view as well. Israel is worse than a mere aggressor. It is viewed as a "colonial" power, guilty of expropriating land that has always been Arab. It is an artificial state, transplanted into the region by foreign settlers.

There are of course a few glaring omissions in this assessment. Rarely is it brought to mind that Jews have maintained a physical presence in the land called Israel for the past three thousand years. Nor is it pointed out that the remains of the civilization they built lie directly beneath the archaeologist's trawl, no matter where we might dig in today's Israel, in Judea and Samaria (the "Palestinian territories"), or even in significant parts of the Hashemite Kingdom of Jordan.

There are multiple places in the world today where the facts we are about to review could never be presented, at least in a public forum. Europe is a prime example. Those who try are surrounded by protesters, booed, heckled and occasionally subject to physical attack. European professors employ anti-Semitic stereotypes, and there are even death threats against those who convey any ideas other than the mantra that

Palestinian Arabs are the true "owners" of the land of Israel. Fortunately, America is not as dominated by political correctness as is most of Europe, at least for the time being. I will therefore unashamedly present this material in the pages that follow and let the detractors rage. The truth about how "Arab land" came to be "Arab land" is not particularly complex. If archaeology be our guide, we may, in a manner of speaking, transport ourselves backward a good three millennia and form some educated opinions about who lived where and when. The story is written in stone, and all one need do is turn over a spade to find it.

Strolling Through Antiquity

How far back in time may we trace the Jewish people, and their biblical forebears, the "Israelites?" Did the Hebrews really come out of Egypt in a mass "exodus?" Did they really take possession of the land of Canaan through the righteous might of their illustrious general, Joshua? Did they occupy the whole territory that came to be called Israel, as the Bible recounts, "from Dan to Beersheba?" These, among many others, are questions for archaeologists, and many are still in dispute. What is not in dispute, however, is that Jews were living in the land in large numbers during the days of the Roman emperors and beyond. How, then, did they manage to become scattered to the "four corners" of the world in a great diaspora that left their ancient homeland largely devoid of Jews, until modern times? What on earth happened? Since most people today have nary a clue, it might well be in order to summarize the story of the Jewish people, step by step.

The ancient kingdom of Israel, at least as ascribed by the Bible to David, Solomon and their progeny, slowly collapsed by the sixth century, before the Common Era (B.C.E.). Having lost their independence to the ancient Babylonians, the people of Israel languished in bitter exile for nearly a century, until a mighty potentate rose in Persia. Cyrus the Great, as he was known, ruled a vast new empire that included the selfsame Israelites (now known as Yehudim – Jews) formerly exiled to Babylon. Now, by imperial edict, they were allowed to return to their conquered land, Judah, rebuild their ruined city, Jerusalem, and their fallen temple. An ancient stone cylinder referencing the progressive policies of Cyrus himself, obliquely bears witness to the events described in the biblical narrative. Archaeology and modern geopolitics came together in that single archaeological relic, given that President Harry Truman, perhaps the most biblically grounded chief executive in American history, referred to himself (upon recognizing the state of Israel in 1948) as the "new Cyrus."

During the centuries that followed, the Jewish people were able to re-establish an independent kingdom of their own, from Judea in the south through Galilee in the north, rivaling in size the earlier Israelite kingdom. Unfortunately for them, this renewed Jewish dynasty was swallowed whole by the truculent legions of Rome in the year 63 B.C.E., under a brutally efficient general named Pompey. For the Jews, this new reality was intolerable. They could think of nothing but revolt, which they undertook in the year 66 of the Common Era (C.E.). Their liberation movement ended in catastrophe four years later, when the Romans, under their general and later emperor, Titus, leveled Jerusalem to the ground, burning their fabled temple with it. That year, 70 C.E., would mark the greatest tragedy in Jewish history, bar none, until the Holocaust.

Even so, the Jewish people survived, and while they never rebuilt their temple, they did rebuild their shattered communities. In the year 132 C.E. they launched a second great revolt, led by a messianic pretender known as Simon Bar Kokhba. Arguably, its results were even more catastrophic than the first, resulting, over the next three years, in nearly six hundred thousand casualties. Jerusalem was sowed with salt, renamed Aelia Capitolina, and Jews were barred from entry, on pain of death, except on one day a year, Yom Kippur, to mourn their fallen shrine. This is how the last remaining vestige of the temple, its western retaining wall, came to be known (somewhat pejoratively) as the "Wailing Wall."

The new Roman emperor, Hadrian, was determined to obliterate every trace of Jewish presence in this rebellious province. Accordingly, the Romans reached back in time to the ancient Philistines, long vanished as a people, but whose heritage supplied the Latin term Palestina – Palestine. The fabric of Jewish society was so shattered that the center of cultural and religious life shifted, from Jerusalem, to the Mediterranean coast, to Galilee. Eventually, it moved still farther eastward, to Babylonia, where, during the centuries to come, the great compendium of Jewish learning, the Talmud, would take its final form. Nevertheless, we should not be so naïve as to think that the Jewish people relinquished their claims on their biblical homeland, or that they completely abandoned it for proverbially "greener pastures." The archaeological record bears out the continued presence of Jews in Palestine, during uninterrupted centuries of Roman rule.

As noted by the British Royal Commission in a 1937 report:

Always ... since the fall of the Jewish state some Jews have been living in Palestine.... Fresh immigrants arrived from time to time ... [and] settled mainly in Galilee, in numerous villages spreading northwards to Lebanon and in the towns of Safad and Tiberias.[1]

Under the "soldier emperors," Antoninus Pius, Marcus Aurelius, Septimius Severus, and Caracalla, the Romans reached what might be called a general "understanding" with their Jewish subjects. Both pagan Romans and sternly monotheistic Israelites would leave each other alone. Twentieth century historian Jacob De Haas observed:

The East breathed more freely and enjoyed even a spell of real peace during the reign of Alexander Severus (222-235).... His predilections brought him the nickname of Archisynagogus, or rabbi. He flirted with the Jews, and his mother, Julia Mammae, protected the great church father, Origen. This catholicity was even exhibited in the imperial palace, where pictures of Orpheus, Jesus, and Abraham, hung side by side.[2]

During this period. the Jewish leadership, its "Patriarchate," was officially recognized by Rome, the result being that Jews remained deeply connected with their homeland. Far away in Babylonia, the great academies of Sura and Pumbeditha were flourishing under a class of civil leaders known as exilarchs, yet the direction of prayer was still to Jerusalem, and the Jewish population of Palestine by no means vanished. Rome gave the Jews official "immunity" from emperor worship, the only subgroup under its authority who were so privileged.

Fate, however, would not be kind. In the year 313 C.E., Rome's mighty potentate, Constantine the Great, issued his widely acclaimed Edict of Milan, legalizing Christianity and opening the door for the full Christianization of the empire. On a practical level, that meant a shift from the toleration of Jews to rabid persecution. The construction of Constantinople, beginning in 324 C.E., would usher in the dawn of the Byzantine Empire, under which Jews and their ancient faith would be pressured and persecuted without mercy. Marriages between Jews and non-Jews were banned. Proselytism to Judaism was forbidden on pain of death. Perhaps because it was recognized that Christianity not only branched off from Judaism but was at its core intrinsically Jewish, it be-

---

1 See Joan Peters, *From Time Immemorial: The Origins of the Arab-Jewish Conflict Over Palestine* (Chicago: JKAP Publications, 2001), 147.

2 See Jacob De Haas, *History of Palestine: The Last Two Thousand Years* (New York: Macmillan, 1934), 147.

came necessary to prove repeatedly that "we are right" and "they are wrong." Whatever the reasoning, the later Roman Empire and the Byzantine Empire that followed essentially determined that Judaism had no right to exist at all. In any case, the Jewish population continued to be large, and Christians were relatively few in number through the fourth century.

A brief respite occurred in 614 C.E. when the Persians took Palestine from its Byzantine suzerains. In his classic tome, *The Jews in their Land*, modern Israel's first prime minister, David Ben-Gurion, noted that large numbers of Palestinian Jews arrived from the hills of Palestine to fight alongside the Persians.[3] It is estimated that between twenty thousand and twenty-six thousand Jews joined the struggle to repulse the Byzantine persecutors.[4] The Persians subsequently turned out to be remarkably pro-Jewish in their policies, to the point that they actually gave the Jews control of the city of Jerusalem. Again we see that the Jewish people never voluntarily abandoned their land, nor did they ever relinquish it to anyone else unless forced to do so at sword point. Ben-Gurion also noted substantial Jewish communities in coastal towns, the vicinity of Nazareth, the Negev desert to the south, and Transjordan. Moreover, when Persia actually handed Jerusalem over to the Jews of Palestine, they were quick to accept. Unfortunately, this sanguine state of affairs would soon be overturned when the Byzantines swarmed in again, under the emperor Heraclius in 627 C.E., General persecution resumed, exacerbated by revenge, in that the Jews had fought on the side of the Persians. The result was a bloody massacre. Nonetheless, historical records indicate that some Jewish settlements remained in their ancient locations, unchanged over millennia. Over the course of time, they were no longer the majority of the population, but they were certainly an important element, staking their claim to the land of Israel as no one else had, from generation to generation.[5]

"Just Showing Up:" The Arab Centuries

During the early seventh century of the Common Era, many Arabian Jews (descendants of those who had fled to the east before the Roman juggernaut) were uprooted when adherents of the new faith, Islam, seized their homes and property. Ironically, they made their way back

---

3 David Ben-Gurion, *The Jews in their Land*, (London: Aldus Books, Ltd., 1974).

4 See James Parkes, *Whose land?: A History of the Peoples of Palestine* (New York: Penguin, 1970), 60.

5 See Bernard Lewis, *The Arabs in History* (Oxford: Oxford Univ. Press, 1967), 49.

to Palestine, where they were united with the Jews who had never left. However, in 634 C.E. they were invaded yet again, by Muslim warriors from the deserts of Arabia – the very land they had just fled. Again they suffered plunder at the hands of their conquerors. Losing their homes and possessions, many were once more turned into refugees. This, more or less, is how the land of Israel went from being Jewish to "officially" Arab. If, as it is said, "eighty percent of success is showing up," then the root of the Muslim-Arab claim on Palestine is the fact that at this particular point in time, they just "showed up."

Historian Philip Hitti observed that they "... brought with them no tradition of learning, no heritage of culture."[6] Indeed, during the course of time, the new Arab overlords became "diluted" by intermarriage with local populations, the end result being a polyglot mixture of ethnicities. The Jews of Palestine, by contrast, remained an integrated community that, even though a minority population-wise, preserved their traditions and their claims to the land. In Jerusalem, Judea, and Galilee, the ruling authorities realized that they were now in charge of a substantial Jewish presence.

Enter Caliph Omar, who would lay the foundations for two great Muslim shrines atop the Temple Mount, the al-Aqsa Mosque and the Dome of the Rock, effectively obliterating the Jewish past on the holy hill. The construction of these buildings is part of a wider practice in the Muslim world of taking over existing sites and landmarks and effectively "Islamizing" them (another example being the conquest of Constantinople and the conversion of Constantine's great basilica, the Hagia Sofia, into a mosque). To the extent that the heart of today's Middle East conflict revolves around the city of Jerusalem and this sacred ground, we must point to Caliph Omar as the instigator *par excellence.*

In his wake came Caliph Omar II, a true zealot for Islam who made sure that all non-Muslims be relegated to the status of *dhimmi* – "protected" people. Such "protection" in this case amounted to permanent second-class status, in which prejudice and humiliation became the rule of law. This likely provided impetus for the rise of a new messianic aspirant known as Serene. He was born a Christian, though according to one early chronicle, proclaimed himself to be Moses, and was "sent again for the salvation of Israel." Serene was instantly hailed by many Jews suffering under the heavy taxation of the Muslims, but he was ultimately arrested and tried before Caliph Yazid II. He recanted of the

---

6 See Philip Hitti, *The Arabs: A Short History* (South Bend, IN: Gateway, 1996), 99.

error of his ways, and his followers returned to "traditional" Jewish life in the synagogue.[7] It was time to settle down for centuries of quietistic obeisance to Muslim rule, waiting on the Messiah. In any case, pockets of Jews tenaciously continued to live in Palestine under the rule of the multiple dynasties and caliphates that followed, from the early Islamic period, through the "interruption" of the Crusades, and down to the present. These included: the Umayyads, the Abbasids, the Fatimids, the Seljuks, the Crusaders, the Ayyubids, the Mamluks, and the Ottomans.

Far to the west Jews lived across the length and breadth of what had formerly been the Roman Empire, but which was now subdivided among warring fiefdoms, including Goths, Visigoths, and Huns. The designation "wandering Jew" was well-earned, for they were constantly subject to attack, plunder and expulsion. They were butchered by marauding Crusaders, en-route to the Holy Land. They were accused of kidnapping Christian children and draining their blood for use in the baking of matzah. They were charged with stealing the consecrated bread of Christian communion rites – the "Host" – and "desecrating" it in mock re-crucifixions of Christ. They were blamed for the Bubonic Plague, which decimated Europe's population in the fourteenth century. (They were presumed to have poisoned the water wells of their Christian neighbors.) In almost perfect century intervals, they were driven from England (1290), France (1394), and Spain (1492). In the same year that Columbus' little fleet set sail in hope of finding a new route to the East Indies, a much larger fleet, laden with at least two hundred thousand Jews, was embarking from Spain for any port that would take them in. The majority found refuge to the east, in Arab lands, some landing, once again, in Palestine.[8]

Looking to Zion

In 1799, Napoleon Bonaparte fought his way from Egypt up the Mediterranean coast, with the intention of creating a new mini-nation, under French hegemony, for the Jewish people. The state of Israel might have been born under Napoleon's suzerainty, had the French conqueror not himself been repulsed by the British at the coastal city of Acco. In any case, the Jewish dream of return to their ancient land would be "born again" later in the nineteenth century, when an assimilated Jewish journalist from Vienna, Theodore Herzl, found himself reporting on a

---

7 See Dan Cohn-Sherbok, *Israel: The History of an Idea* (London: SPCK, 1992), 67.

8 See Jonathan S. Ray, *After Expulsion: 1492 and the Making of Sephardic Jewry* (New York: New York University Press, 2013).

new outrage in Paris. A Jewish officer in the French army, Captain Alfred Dreyfus, had been charged with passing a classified artillery manual to a German military attache. Accused of treason, Dreyfus was sent to Devil's Island, off the coast of French Guiana, where he languished at hard labor. When Herzl heard cries in the streets of "Death to Jews," he experienced an epiphany of sorts. He became aware for the first time that the Jews of Europe were no more secure now than during the long centuries of their dispersion. The only solution would be a sovereign state of their own, preferably in Palestine.[9]

Gathering Jewish delegates from all across Europe, Herzl became the undisputed father of modern Zionism. His guiding principle and adage: "If you will it, it is not a dream." As the Ottoman Turks, who owned Palestine in those days, were not interested in relinquishing their territory, Herzl never lived to see his dream realized. That would take additional decades of blood, sweat, and tears.

It is noteworthy that the Ottoman empire had actually encouraged Jews to settle in the land, especially after their expulsion from Spain. Some of this flock of refugees arrived in a town in northern Galilee called Safed, which became the locus of Jewish mysticism in the sixteenth century and beyond. Later, inspired by Herzl's Zionism, substantial numbers of Jews began arriving in Ottoman Palestine, mostly from eastern Europe and Russia, to lay the foundations of what would, in the year 1948, become the modern state of Israel. Notwithstanding the refusal of the Ottomans to sell the land to Jewish settlers, the Jews would create "facts on the ground."

The goal of this "practical Zionism" was to create a Jewish majority in Palestine, for the first time since late antiquity. After all, if the mere "showing up" of Muslim invaders, sans any historical/cultural link to the land, was enough to make it "Arab territory," then why should the "showing up" of large numbers of European Zionists in the late nineteenth and early twentieth centuries not be enough to make it "Jewish territory?" If we wish to go back in history, why not go "all the way" back, to discover exactly how much Jewish civilization lies beneath the surface of this Palestinian Arab "homeland?"

Moreover, it is worth mentioning that when the early Zionists were first arriving and legally buying land, often from absentee landlords living in Istanbul, the Muslim Arab population of Palestine was by no

---

9 For an overview of the Zionist movement see Walter Laqueur, *A History of Zionism: From the French Revolution to the Establishment of the Establishment of the State of Israel* (New York: Schocken Books, 1972).

means what it grew to be in the decades that followed. Demographers have noted that between the years 1919 and 1939, improving job opportunities – the direct result of Jewish efforts to reclaim the land – resulted in an influx of some fifty thousand Arabs from nearby countries.[10] Due to improving living standards and health conditions, the Arab population continued to grow exponentially throughout the period of British rule (the British Mandate), from 1922 onward, swelling to 1.2 million in the whole of western Palestine. However, does this degree of population growth in the twentieth century justify the claim that all of Palestine was historically "Arab land?" What we see instead are "see-saw" population shifts that tell us little, if anything, about the "rightful owners" of the land.

When the state of Israel was born in 1948, the Jewish population in Palestine amounted to around 600,000, substantially less than the Arabs. For this reason, the borders for the new Jewish state drawn by the U.N. included only those areas where Jews were in a clear majority. Over time, however, the Jewish population in the area grew even more, especially as Jews in Arab lands were so oppressed that they had little choice but to flee to the new state of Israel. Today, Jews are the majority, not only within the 1947 boundaries set by the U.N., not only within the expanded borders that prevailed from the 1949 armistice (following Israel's War of Independence) until the Six-Day War of 1967, but in the whole of western Palestine, from the Mediterranean Sea to the Jordan River Valley.

The reality could not be further removed from the incessant propaganda drumbeat that depicts a Jewish minority ruling an Arab majority in an "apartheid" fashion. Current statistics instead reveal that Jews constitute some 58.6 percent of the entire population west of the Jordan River. Arabs comprise not more than 38.7 percent, while other minorities make up 2.7 percent of the total population.[11] Furthermore, birthrates of Arabs in Israel, Gaza, Judea and Samaria are trending downward, while Jewish birthrates are rising. Add to this the fact that Arab immigration rates are in negative territory, while Jewish immigration remains consistently high, and the "apartheid" slander is revealed as nothing more than a myth.[12]

---

10   See Martin Gilbert, *The Routledge Atlas of the Arab-Israeli Conflict* (New York: Routledge, 2005), 16.

11   See Yakov Faitelson, "Demographic Trends in the Land of Israel (1800-2007)," *Institute for Zionist Strategies*, January 15, 2011.

12   See Caroline Glick, *The Israeli Solution: A One-State Plan for Peace in the Middle*

It is also worth noting that well before Jews came to comprise a majority of the population in the region, a plan was put forward to create a Jewish "national home" in Palestine. In 1917 the British government issued the much celebrated Balfour Declaration, declaring:

> His Majesty's Government view with favour the establishment in Palestine of a national home for the Jewish people, and will use their best endeavours to facilitate the achievement of this object, it being clearly understood that nothing shall be done which may prejudice the civil and religious rights of existing non-Jewish communities in Palestine, or the rights and political status enjoyed by Jews in any other country.

We should bear in mind that Palestine in those days comprised not only the entire west bank of the Jordan River, but the east bank as well, including the modern Kingdom of Jordan. Had a modern Jewish nation been created along those lines, it would indeed have approximated the boundaries of the ancient Israelite kingdom that lies buried beneath the surface. Unfortunately for Zionist aspirations, the Balfour Declaration was never fulfilled. Instead, "Palestine" was effectively reduced in size, when Winston Churchill, at a conference in Cairo in 1921, engineered an Arab emirate of Transjordan to the east of the Jordan River, fulfilling the spirit of Britain's promises to the Arabs during the First World War. The throne of the new Hashemite Kingdom of Jordan would be occupied by King Abdullah I.[13] He would preside over what arguably was and yet remains the "Palestinian state." That still left all of the territory to the west of the Jordan as the "national home" envisioned by Lord Balfour.

Nevertheless, no Jewish state ever came about under the British. On the contrary, they choked off Jewish immigration to Palestine at the very time it was desperately needed – on the eve of World War II. After the defeat of the Axis powers, Britain essentially washed its hands of its mandate, declaring its intention to withdraw from the region in May 1948. Consequently, when the U.N. voted to "partition" Palestine in 1947, between a Jewish and an Arab state, it was essentially "re-partitioning" Palestine, into two sovereign nations, neither of which would be geographically contiguous, much less defensible.

Given this background, many Israelis ask: Why must the Jewish state surrender to the Arabs the one commodity of which it has precious

---

East (New York: Crown Forum, 2014), 127-129.

13 See Ronen Yitzhak, *Abdullah Al-Tall, Arab Legion Officer: Arab Nationalism and Opposition to the Hashemite Regime* (Brighton: Sussex Academic Press, 2012).

little left – land? Perhaps it is the Arabs who, based on the reality of history and confirmed by the archaeological record, should be ceding land to the Jews. For many Jewish settlers in Judea and Samaria, known in today's politically correct nomenclature as the "West Bank," the "real Israel" is the underground Israel – whose artifacts lie in great abundance, just beneath the surface. The archaeological remains outline what is often called the "Greater Israel," a dream, to be sure, but an ever-present reality in the minds of a troupe of Zionist zealots, who are not inclined to surrender to a cacophony of voices demanding the step-by-step dismantling of the Jewish state.

A Fictional History Is Born

Thus, we have a brief overview of Jewish life, substantiated, as we shall see, by physical remains in the land of Israel. It is more than odd, however, that modern Palestinians have created a kind of "fictional history," in which they and only they had been living in Judea and Samaria/ "Palestine," as it were, from time immemorial. This, doubtless, is why the Palestinian Authority finds it necessary to minimize, not only the history thus far reviewed but also the physical evidence of Jewish presence in the land itself. The lesson is clear: he who controls the past controls the present and the future. Biblical archeology for its part is hardly a dry academic study of ancient stones, bones, and potshards; it has become a weapon of war.

Archaeologists are of course loathe to admit that their discipline might be co-opted as a pawn in a larger political chess game. Archaeology supposedly has no axe to grind, no hidden agenda. The story told by physical remains is presumably without prejudice or ulterior motive. At least in theory, archaeology is not to be used to justify contemporary claims on land or territory, or to settle the issue of who rightfully lives in a certain locale today. However, when one side of the Arab-Israeli conflict (the Palestinian Authority) attempts to nullify the other's existence by obscuring the archaeological record, all bets are off. Archaeology is a part of today's Middle East conflict because the Palestinian Arabs have made it so.

In creating and perpetuating their own fictional history, the Arabs have cast aspersions on all traces of Jewish presence in their ancient homeland. Some go as far as to ask: how do we know that the Jews have not created their own fictional history, imagining a great Solomonic temple that never existed? How do we know that other Jewish sites, from Rachel's Tomb near Bethlehem to the Tomb of the Patriarchs in Hebron, are not "bogus" as well? Such issues are more germane to modern geo-politics than most people care to imagine. Moreover, they are the exact focal point at which archaeology and the Arab-Israeli conflict come to-gether in ongoing tension. A few examples are called to mind.

In 2011 the Palestinian Authority issued a formal complaint, via its Egyptian representative, Barakat al-Farra, against the Shanghai Expo, in China, for referencing, among other things, the Jewish history of Jeru-salem. The Palestinians also demanded that the United Nations Educa-tional, Scientific and Cultural Organization (UNESCO) recognize the Tomb of Rachel and Hebron's Tomb of the Patriarchs, not as Jewish, but Palestinian sites. Not surprisingly, UNESCO bowed to the pressure. Additionally, on the Information Ministry's page of the Palestinian Au-thority's website there appeared in November 2010 a paper by Ministry official Al-Mutawakel Taha, dismissing any linkage between the Jewish people and the Temple Mount, including the holiest site in Judaism, the Western Wall.[14]

During the Camp David negotiations of 2000, between Israeli Prime Minister Ehud Barak and Yasser Arafat, hosted by Bill Clinton, Amer-ican chief negotiator Dennis Ross was shocked to hear the Palestinian leader declare, "The temple never existed in Jerusalem, but rather in Nablus." In a similar vein, senior Palestinian negotiator Saeb Erekat in-sisted that the "Jerusalem temple is a Jewish invention." A flabbergasted Clinton responded, "Not only do all of the world's Jews believe that the Temple was located on the Temple Mount, but most Christians believe it, too."[15] It was a pivotal moment in the negotiations, for the Israelis instantly realized that the Palestinians had no intention of agreeing to a peace deal of any kind. Polling data showed that ninety-one percent of the Israeli public would reject any outcome resulting in Palestinian control of the sacred plateau. Since this revelation, even the bulk of the Israeli left came to conclude that a real "peace partner" on the Palestin-ian side does not exist. It has even been surmised that if the Palestinians had given voice to such an opinion as early as 1993, the famed Oslo Accords that formally set in motion the principles for the creation of a Palestinian state, would never have gotten off the ground.

All of this highlights the fact that the ongoing conflict and the seem-ingly endless stalemate that is the Middle East today is very much linked,

---

14 See Yitzhak Reiter, "King Solomon's Vanishing Temple," in *The American Interest*, March-April, 2011.

15 Yitzhak Reiter, *Jerusalem and Its Role in Islamic Solidarity* (New York: Palgrave Macmillan, 2008), 39.

not only to historical remembrance, but to what the spade of the archaeologist may or may not uncover. All of this should be viewed against
the growing movement afoot among nationalist, so-called "right-wing"
Israelis, to settle the territories won in 1967. To them Judea and Samaria
are not the "West Bank," but the heartland of biblical Israel. They remain unperturbed by various proclamations of international law, forbidding the settlement of a country's citizenry on territory conquered
and subsequently "occupied." They are equally unphased by the fact that
the activity of the half-million Israelis who live over the "Green Line," in
the conquered territories, is hugely controversial within Israel itself. In
spite of the rage generated, both inside and outside of Israel, "religious
nationalist" settlers represent an increasingly influential segment of the
population of the Jewish state. This includes prominent political and
military figures. Some forty percent of the Israel Defense Force have
indicated that they share the ideology of the "sanctity of the land." On a
demographic level, the settler population surged by over three percent
in 2019 alone, significantly above that of the rest of Israel proper.[16]

It is important to stress that today's settlers are not merely talking in
general terms about the long Jewish presence in the land. They believe
that they are in Judea and Samaria to live literally on top of specific sites
inhabited by their ancestors. The fact that these ancient sites exist will
always inspire the settlers to remember "what we used to be" and to fuel
a mystical longing, perhaps in the Messianic Age or perhaps sooner, to
reestablish a land of truly biblical proportions. Through it all, we are
constantly brought back to the archaeological issue of what this biblical land must have looked like. What exactly were its "biblical proportions?" Will the discovery of additional physical remains add more fuel
to the settlers' fire? Or will the archaeologists on the "minimalist" side
of the debate actually dampen their fervor? One thing is certain. By attempting to invalidate the biblical past, the Palestinian side has actually
opened the door to this debate. It is now incumbent upon us to find out
just how accurate, if at all, the biblical record is.

Rage of the Radicals

There is one more angle to the activity and ideology of certain Israeli
setters and their allies, namely, the "radical" faction, whose aims are by
no means to be vindicated in these pages. Ever mindful that important
Jewish sites have been effectively confiscated and "converted" to Islam,

---

16 "West Bank Settlements Report Rapid Growth in 2019," *Times of Israel*, Jan. 28,
2020.

they have determined in recent decades to do something about it. If a Muslim-Arab shrine stands on an ancient Jewish site, why not "eliminate" it? Why not take back all Jewish sites, by force of arms if necessary? A contemporary Israeli filmmaker even produced a major motion picture about a charismatic West Bank rabbi and a fanatical conspiracy to dynamite the Dome of the Rock from the maze of ancient tunnels underneath. The most frightening aspect of the film is that it represents a fictionalized reworking of a plot that was all too real. A prominent member of a kindred ultra-nationalist group, the "Temple Mount Faithful," went on a tour of the United States, speaking in churches (not being welcome in synagogues), collecting money and declaring, "I would like to take the Dome of the Rock and mail it back to Mecca in little pieces!"

Truth be told, violence and terrorism are not the exclusive domain of the Palestinian Arabs. The security service of the state of Israel, the Shin Bet, has long been waging an internal battle against radical elements among the ultra-nationalists, ironically, to protect Arab holy places, as well as the Arabs themselves. They have broken up active terrorist plots, which were of such an incendiary nature that, if they had succeeded, might have plunged the Middle East into a new and terrible apocalyptic war.

The so-called "Jewish Underground" doubtless looks to great biblical heroes, such as Joshua, who first conquered the fortress city of Jericho and then took one Canaanite city after another, until the whole land, from Dan in the north to Beersheba in the south, lay at the feet of the Israelites. What Joshua did over three millennia ago, they are prepared to replicate. Beyond the geopolitical implications of such bravado, the task of the archaeologist is to test every biblical assertion with good science. Indeed, we must not shy away from addressing serious objections to the Bible's "lay of the land."

Was there really a conquest of Jericho, as the book of Joshua declares, along with a fallen wall? Was there a great Israelite conquest of Canaan? Was there even a Joshua, a Moses, a David or a Solomon? What if the biblical account of conquest is in the end found wanting? How much exaggeration might there have been in the description of the Israelite kingdom of David and Solomon? Would any of that make a difference to Israel's homespun "radicals?" Obviously, we have much to tackle in the pages that follow. We may as well get under way.

CHAPTER ONE

SETTLING HISTORY –
MOVING IN ON DAVID'S CITY

Jerusalem, July 2011 C.E.

A TINY BELL. A tiny golden bell. Might this be the "signal" for which a small sect of faithful Israelites have waited for two millennia, declaring that the time is ripe for rebuilding the ruined temple of Jerusalem? By contrast might it be a "provocation," as Palestinian sources claim, in the ongoing conflict over who owns the most disputed piece of real estate on earth? According to Palestinian television news, "Israeli sources said that a gold bell had been found in a tunnel, belonging to what they call the Second Temple Period. This once again highlights the efforts of the Israeli side to forge the history of the holy city."

Forge history? Is that the sinister campaign Israel's biblical archaeologists are pursuing? Has a dark conspiracy been hatched, to take control of Muslim holy sites, occupying Islamic holy ground, and build in their place a new Jewish temple? If so, World War III might be expected to arrive much sooner than previously imagined. In any case, there can be no doubt that archaeology is becoming one more weapon in today's ongoing Arab-Israeli conflict.

How could a little bell produce such rancor, when its discovery should have been met with awe and wonder? Its design is exquisite, perfectly preserved and untarnished after two thousand years. Found in the remains of a drainage tunnel dating to the first century of the Common Era (the entrance of which lay just west of the famed Western Wall of the Temple Mount), the excavation managers declared, "It appears the bell was sewn on the garment of a man of high standing in Jerusalem at the end of the Second Temple period."[1] The Bible describes the bluish tunic of the temple's high priest as being festooned along its hem with seventy-two gold bells, interspersed with woven pomegranates. Given the location of this find and the fact that it perfectly matches the biblical description, it is not beyond the realm of possibility that this minuscule artifact may have been worn by the man who presided over ancient Jerusalem's "Holy House." Even the miniature clapper remains in place, and when shaken, it still produces a faint ringing sound. In short order, audio files began appearing on the internet, beckoning the listener with "a voice message from the holy temple."[2]

Jerusalem Map, 1874

1   Attributed to Eli Shukrun and Ronny Reich; see Hana Levi Julian, "Sound of the Kohen Gadol's Golden Bell Rings Again in Jerusalem," *Jewish Press*, Oct. 10, 2016: https://www.jewishpress.com/news/breaking-news/sound-of-the-kohen-gadols-golden-bell-rings-again-in-jerusalem/2016/10/10/.

2   https://www.youtube.com/watch?v=vTN-uEsAAaY.

David's City: Defying the Propaganda

As amazing a find as it is, the little gold bell is just the most recent episode in a long saga of archeological discoveries that have confirmed the biblical descriptions of Jewish history. We are about to travel back in time and explore the most significant among them, along with their discoverers, and with each one of them, we will also present not only the "facts on the ground," but the science that underlies them and the inevitable criticism from the Palestinians and their advocates in the media. Sadly, in today's geopolitical climate, such voices are as much a part of the story as the discoveries themselves. A suitable place to begin is in a sleepy suburb in East Jerusalem, the name and significance of which is of little interest to most – Silwan.

The village of Silwan, c. 1865

With a population of some forty thousand Arabs, it is, given the tensions of the region, much less sleepy and much more confrontational than might be imagined. This is because Silwan is the location of a major archaeological site, advertised as the City of David (Ir David in Hebrew). It is part of a natural rocky outcropping extending southward from a ridge-like area known as the Ophel, that in turn connects with the south end of the Temple Mount. Long ago, in the twelfth century B.C.E., there was a city here called Jebus, after the Jebusite tribe of Canaanites – those conquered by the biblical hero Joshua. According to the Bible, the city remained in Jebusite hands until it was taken by King David, somewhere around the tenth century B.C.E. Such was the genesis of the city that is today known as Jerusalem.

All of this, however, is incessantly undermined by Palestinian propagandists. David and his memory are minimized, if not erased from history. Moreover, excuses are being made for the apparent lack of Palestinian archaeology, while Jewish archaeology proceeds at fever-pitch. In a television interview for Al-Jazeera, archaeologist Mahmoud Hawari of Bir Zeit University declared:

> The reason Palestinians are not as interested in archaeology as Israelis is because we don't have to prove that we were here two thousand years ago.[3]

Put simply, it is so obvious that Palestinian Arabs were living, some two millennia ago, in the land that today comprises the Jewish state, that they are disinterested in archaeological attempts to prove it. One might, however, suggest another scenario. It is painfully evident that today's Palestinian Arabs can trace their residence in the land no further back than the Muslim conquest of 634 of the Common Era. The great majority arrived only within the last century or so. To link themselves to the Nabatean Arabs who lived there in antiquity, and who were, after all, pagans, not Muslims, is at best a vain exercise. Could it be that modern Palestinians simply have nothing that old for which to dig?

Furthermore, Palestinian detractors make a more damning accusation, maintaining that Silwan has been overwhelmed with gun-brandishing Jewish settlers, who seek to annex it for Israel. In a television segment on the City of David, produced for the CBS television program *60 Minutes*, correspondent Lesley Stahl observed that "the challenge is how to divide the city between the two sides." The underlying assumption is of course that the city must indeed be divided. The trouble with such political correctness is that it is repeated so often that target populations sheepishly give assent. Counter arguments are ignored, and facts are no longer at issue.

In reality, East Jerusalem does not need to be claimed by Israel; it is already part of the Jewish state, and has been since 1967. It was formally annexed by Israel's "Jerusalem Law" of 1980, which declared, "Jerusalem, complete and united, is the capital of Israel."[4] Those are the "facts on the ground." Wars, like elections, have consequences. Israel did not simply arrive one day and seize Silwan. Israel was attacked in June 1967

---

3 Al-Jazeera, "Looting the Holy Land," 22:15.

4 https://www.knesset.gov.il/laws/special/eng/basic10_eng.htm.

by the Hashemite Kingdom of Jordan, following its preemptive strike on Egypt. In spite of the fact that the Israeli government cabled King Hussein, urging him to stay out of the conflict, artillery fire was initiated along Jordan's entire border with Israel. The fighting in Jerusalem was particularly brutal. In the end, however, East Jerusalem, including the Temple Mount, and sleepy Silwan, fell to the Israelis. This was a defensive war for Israel's survival, and taking Silwan was matter of securing victory, not initiating an "illegal occupation."

Silwan excavation

The David Dig Goes Political

Today, much of Silwan is being turned into an archaeological park. Roughly 400,000 tourists visit the site each year, with the opportunity to trudge through the flowing waters carrying small flashlights – part of the admission price – all the way to its exit, inside the Old City walls, at the Pool of Siloam, where Jesus was said to have healed a blind man. The rancor from the Palestinian Arabs and their allies in the international media is predictably vicious. This entire site amounts to a commercialization of archaeology for the nefarious purpose of establishing an Israeli presence in Silwan and annexing it to Jewish Jerusalem. After all, the official logo of the City of David is a harp – very sinister indeed. David's harp happens to be the logo of Ir David, allegedly an Israeli settlers' organization. Is this not proof that the City of David archaeological park is a conspiratorial endeavor, backed by the very Israelis who are building Jewish settlements in the "occupied territories?" It is in fact the case that the City of David is administered by the Ir David Foundation (also called El'Ad), and this is the point at which the controversy becomes heated.

We should consider the accusations leveled against the Israelis on account of David's ancient city itself. It is alleged that the Israelis were intent on incorporating the village of Silwan into Jewish Jerusalem ever since the city was unified ("illegally conquered," in Palestinian parlance), in 1967. In the mid-1980s, when a commander of an elite unit of the Israel Defense Force, David Be'eri, visited this part of greater Jerusalem, he found that the specific locations of past digs were covered with garbage and waste. He subsequently left the military to establish the Ir David (or "City of David") Foundation, commonly known as El'Ad, a Hebrew acronym which means "to the city of David."[5] Its stated purpose was to retrieve and restore the city's ancient heritage. To advance this goal, the El'Ad Foundation helped fund a new excavation at the summit of the City of David, headed by Eilat Mazar, on behalf of the Institute of Archaeology of Hebrew University of Jerusalem. The Israel Antiquities Authority oversaw the excavation, but the site itself was mostly funded and operated by El'Ad.

Also in the 1980s, Israeli Prime Minister Ariel Sharon's housing ministry is alleged to have plotted to take over Arab properties in Silwan, as well as the Old City to the north, by classifying them "absentee properties" – having originally been built without permits. Sharon's modus-operandi was clear enough, namely, to create more "facts on the ground," to match the ones underneath. Jews were said to have seized property under the guise of the "Absentee Property Law," and via indirect land sales, even while still occupied. Jordan's permanent representative to the U.N. responded in 1987 with a formal complaint about Jewish settlement activity. His letter alleged that the Israelis took possession

---

5   See Margaret Haerens, *Suicide* (Detroit: Greenhaven Press, 2012), 156-9.

of two Palestinian homes in the adjacent neighborhood of Al-Bustan, having ejected their occupants.

That, according to the Palestinian Arabs, was only the beginning. The Ir David Foundation was accused of consorting with another Zionist movement called Ateret Cohanim ("Crown of the Priests") to spur ongoing Jewish settlement in East Jerusalem. The latter undertook construction in 2003 of a seven-story development called Beit Yonatan, after convicted Jewish American spy, Jonathan Pollard, even though they lacked proper permitting. In 2007, the Israeli courts ordered that Beit Yonatan be evacuated and sealed. Jerusalem's mayor, Nir Barkat, stepped up, arguing that this Jewish housing project should be treated in the same way as Arab housing, much of which was approved retroactively. The "no double standard" argument won the day, and the residents of Beit Yonatan, some four hundred fifty strong, remain in their flats, though admittedly under heavy security and surrounded by neighbors who are, to say the least, hostile.[6]

An additional accusation was that the Jewish National Fund drew up protected tenant agreements with Ir David, by which the foundation began construction without going through the approval process. As a result, over fifty Jewish families moved in, some living in houses acquired from Arabs, who had no idea that they were selling to Jews. Fifty Jewish families among tens of thousands of Arabs was deemed a disgraceful intrusion on Arab land. Subsequently, one hundred thirty illegal Jewish structures were reported in 2009 by the State Comptroller's Office, an increase of thirty in 1995 and eighty in 2004. Not surprisingly, Palestinian sources engaged in a bit of hyperbole in responding to such a "provocation:"

> It should be noted that the decisions to destroy and remove the Al-Bustan neighborhood were meant to realize the Judaization plans, according to which Talmudic parks would be established in its stead, serving the myth and legend of the *alleged* temple.[7]

The reality was, as might be imagined, somewhat different. A case can be made that it is the Palestinian Arabs who have for the most part been illegal squatters in Silwan over a number of decades. Ever since the the land was ruled by the Ottoman Turks, and later under the British

---

6 See Ari Soffer, "Who are the Jews of Silwan and Why are They There?," *Arutz Sheva*, Oct. 22, 2014: http://www.israelnationalnews.com/News/News.aspx/186436.

7 *Al-Hayat Al-Jadida*, May 12, 2011.

Mandate, the lower part of Silwan was wooded and building-less, so as to preserve its unique historical character. However, during the last two decades alone, Arab squatters built some ninety houses in the section of Silwan known as King's Garden, without permits. It might be perceived as at least somewhat ironic that Palestinian Arab illegal squatters are claiming that Israeli Jews are ... illegal squatters. Israel's decision was to approve sixty-six of the ninety Palestinian houses retroactively, while granting alternate plots of land to the others, who were to be resettled.

As the conflict continued, however, so did the invective, the Jewish state of course being the recipient, even though the dig was in "private" hands. The *60 Minutes* piece featured a Palestinian "activist" named Jawad Siyam, who claimed that he could trace his family roots in East Jerusalem back nine hundred thirty years. The cameraman dutifully followed him as he organized a demonstration, culminating in a shouting match with one of the Israeli staff members working at the site. Angry that El'Ad bought his grandmother's house, he shouted:

> You will be the rubbish of history! There is no proof that King David was here! You want to take our land![8]

Queried about a possible "two state solution," by which Israel and the new state of Palestine would coexist side-by-side, he responded:

> I don't think there will be a two state solution. It's not possible to do it. Today, the settler groups are much stronger than before.[9]

If he in fact believed a two state solution to be impossible, why was he demonstrating? What does he mean by his "prophecy" that Israel will be "the rubbish of history?" What does he want? Was he advocating a "one state solution," in which the state of Israel simply vanishes? Such a follow-up question was of course never asked.

To be sure, Israel, unlike its Arab neighbors, is an open, pluralistic society, which sometimes makes the Jewish state its own worst enemy. The country is divided on virtually every subject, from strategic defense matters, to economic policy, to the daily weather report. It is not surprising that Israel has a sizable and vocal left-wing bloc that is so concerned with "fairness" and "social justice" that it occasionally sides with the Pal-

---

8   http://www.cbsnews.com/stories/2011/06/05/60minutes/main20066909_page2.shtml.

9   Ibid.

estinian Arabs. This of course plays into the agenda of the international media, ever-willing to demonize Israel. Palestinians, it is said, are being forced out of their homes. It is irrelevant that these homes were built on state-owned land and lack proper infrastructure. It is equally irrelevant that the municipality of Jerusalem has retroactively approved two-thirds of the homes in any case, and must now install suitable plumbing and electricity. It is also irrelevant that the rest of the squatters will be given land of equal value in spite of having violated the law. Detractors on the left never tire of excoriating Israel while siding with the Palestinian Arabs, perceived in the world media as "oppressed."

Private Sector Archaeology vs. "Big Government"

As for the archaeology itself, the Palestinian Arabs charge that many of the finds amount to nothing more than distorting science to establish dubious "biblical roots." They have even found some unlikely allies among the Israeli archaeological community, such as Tel Aviv University's Raphael Greenberg, who declares, "Archaeologists have given up many of their best practices in order to answer the continuing demands of mainly political actors."[10] He observes that certain archaeologists are in the pockets of financial donors, such as the Ir David Foundation, in trying to authenticate a biblical heritage. This in turn increases tourism and solidifies the Jewish hold on the area. Says Greenberg, "Over time, when you're funded by these people in huge sums, and we're talking millions of dollars, you become part of the machine."[11]

Let us assume that such criticism has merit, and that it would have been better if an arm of the Israeli government were to administer the City of David, rather than a private ("settlers") organization. It should in any case be conceded that things in the "real world" rarely correlate with ideal scenarios. If the government of Israel had been in charge, would the excavations have proceeded apace, or would the Israeli government have done what governments always do — delay? Moreover, since Silwan is "contested" territory that might one day become part of a new Palestinian state, might the entire endeavor simply have to come to a halt? We need hardly be reminded that governmental operations never seem to proceed smoothly, whereas the private sector usually manages to accomplish things efficiently and in a timely manner, leaving gov-

---

10   Erika Solomon, "Researchers dig up controversy in Jerusalem," Reuters, March 24, 2010: https://www.reuters.com/article/us-palestininians-israel-jerusalem-digs/researchers-dig-up-controversy-in-jerusalem-idUSTRE62N33D20100324.

11   Ibid.

ernment projects hopelessly behind, mired in bureaucracy, and playing catch-up? The excavations at the City of David have in fact been moving ahead at such a fevered pace that detractors are complaining of cave-ins in the vicinity. While this may sound like a serious deficit in the work of the excavators, Israeli Professor Israel Finkelstein notes that the excavations themselves are being carried out lawfully and according to competent archaeological methodology.[12]

Nonetheless, media reporters, looking for a scoop, must search for controversy, dredging it up if necessary. When pressed on the question of motive during the *60 Minutes* piece, El'Ad's International Director of Development, Doron Spielman, observed with some pride that "It's all about archaeology, and rebuilding a Jewish neighborhood." He also admitted that it is about "buying homes and buying land." Why is buying homes and land such an outrage? Might it be that it is because Jews are the buyers? Says Mr. Spielman, "If there's a home that an Arab wants to sell, and I have the money to buy it, and I can enable a Jewish family to live there, and I can dig archaeologically underneath it, then that's a wonderful thing to do."[13]

CBS News points out that tens of millions of dollars have come from the United States to this end, as if it is somehow shameful for individual Americans to donate money privately to a Jewish organization in Israel. The report repeatedly stresses that this is land "claimed" by Palestinians for their future state. Implicit in this line of reasoning is that any area claimed by Palestinians must be kept (to employ a vicious Hitlerism) *Judenrein* ("free of Jews") in perpetuity. CBS's agenda appears to have much in common with the Al-Jazeera broadcast, the narrator of which states matter-of-factly, "A number of militant Jewish settlers have moved into Silwan to claim the village for Israel." Are we then to assume that every Jew who manages to purchase property in territory claimed by Arabs is by definition "militant?" Does this include the Jewish "settler" woman interviewed by CBS, Deborah Adler, who related, "We see ourselves as everyday regular people, living in a very, very special place?" It is true that some Israelis living in Arab neighborhoods, including the residents of Beit Yonatan, carry weapons, but are *they* the militant ones, or is it their Palestinian neighbors, who might otherwise murder them?

May we be so bold as to make a correlation with migration patterns

12   Israel Finkelstein, "'Looting the Holy Land' or Pillaging the Truth?," *The Bible and Interpretation*, Dec., 2010: http://www.bibleinterp.com/review/mov2.shtml.

13   *60 Minutes*, "Controversy in Jerusalem: The City of David," June 5, 2011: http://www.cbsnews.com/stories/2011/06/05/60minutes/main20066909_page2.shtml.

in other parts of the world? Is there similar outrage about the so-called "Japanese invasion" of the Hawaiian Islands by tourists and businessmen, with the result that such quintessentially American hotels as the Hawaiian Regent, the Surfsider and the Imperial Hawaiian are now owned by foreign interests from Japan? While some angst about this has developed in certain circles, who would seek to prohibit private individuals from whatever country from buying property wherever they like, as long as the owners are prepared to sell? Israeli Jews, however, must not buy land from Arabs, especially if it involves establishing the existence of the Bible's greatest king.

If or when a Palestinian state finally comes into existence, and the village of Silwan becomes part of it, what will be the status of the City of David? Will there be open and unfettered access for Israelis and Jews from abroad, who simply wish to visit the capital of the greatest biblical hero of all? Will excavations be allowed to proceed at all? As with everything in the Middle East, double standards are pandemic. When Israel conquered the Temple Mount (known in Arabic as the *Haram al-Sharif*, the "Noble Sanctuary") in 1967, did it impede access to the Muslim shrines on top of the massive platform? On the contrary, in creating the *Waqf* ("Islamic Trust"), which oversees the entire area, it acceded to Islamic authority in that part of Jerusalem, including subterranean "renovations" beneath the al-Aqsa Mosque that have met with no interference, in spite of the fact that they have been carried out with no regard whatsoever for proper archaeological methodology. The Muslim authorities are not pursuing archaeology in this case; they are destroying it.

When it comes to David's palace, there is certainly a good deal more history and archaeology behind it being located just south of the Ophel than ever there was about Muhammad's night ride from Mecca to Jerusalem, on his mystical steed al-Buraq. Are we in fact asked to believe that the prophet dismounted al-Buraq, set foot on the very rock over which the golden dome stands today, and ascended into heaven on a ladder set up by the angels Jibril (Gabriel) and Mikhael, for an audience with Allah? Political correctness nonetheless cautions against casting aspersions on Islamic tradition. When it comes to Jewish/ biblical traditions, however, it is generally "open season."

One of the Jewish "militants" of the City of David, featured in the *60 Minutes* piece, Yonatan Adler, is unworried. He insists:

The City of David is where Jerusalem began. This is where prophets walked. This is where half of the Bible was written.... Maps are written on paper; this is written on our hearts.

Lesley Stahl is not so sure: "There's actually no evidence of David, right?" At this point one is inclined to wonder who in fact is "right."

The Facts Underground

Sherlock Holmes, in Conan-Doyle's "A Scandal in Bohemia," famously criticized those who "twist facts to suit theories, instead of theories to suit facts." In this case, the facts underground, now being excavated after millennia of dusty silence, are seen by Israelis, "settlers" and ordinary citizens alike, as one more verification that this land is theirs. To what extent is serious "twisting" taking place here? One might put Doron Spielman in that category when he declared that Abraham no doubt had laid eyes on this very site. To be sure, there is nothing about Abraham, who lived nearly four millennia go, according to the Bible, that can be proved archaeologically.

Nevertheless, the underground facts at the City of David are multiple. In the nineteenth century, famed British archaeologist Charles Warren arrived at the shocking conclusion that the "real" City of David lay outside and well to the south of the medieval city walls seen today. In October 1867, the inveterate explorer made his way through a water conduit leading away from the natural spring to the east of the city called the Gihon. In an Indiana Jones-type adventure, Warren and his team crawled hundreds of feet into this tunnel, occasionally up to their mouths in the icy water. At nine-hundred feet in, they came upon a series of false turns and found themselves in a zig-zagging pattern. After nearly four hours in the cold spring water, they came out, shivering in the darkness, convinced they had found David's access to Jebus, which he subsequently captured.[14]

Today, this water conduit is known, not as David's, but as Hezekiah's Tunnel. We now know that it was built by the biblical king who lived two centuries after David. In trying to defend Jerusalem against the Assyrian general named Sennacherib, who was about to besiege the city, Hezekiah produced a masterpiece of ancient engineering that helps define the City of David and the village of Silwan, under which it courses to this day. This is the tunnel that flashlight-carrying visitors to the

---

14 See Sir Charles Warren, *Claude Reignier Conder, The Survey of Western Palestine: Jerusalem* (London: Alexander P. Watt, 1889), 367.

City of David still traverse, thankfully never more than waist-deep. An ancient inscription in paleo-Hebrew, found halfway through the tunnel and today residing in the Istanbul Archaeology Museum, commemorates where the ancient diggers, coming from each end, came together. The inscription, like the tunnel as a whole, can hardly verify every detail of the biblical story, but it is yet another fact underground, testifying to Israelite presence in exactly this part of ancient Jerusalem.

Hezekiah's Tunnel and inscription

This set of facts of course relates to an Israelite king of the eighth century B.C.E. What can be known about David, of the tenth century, in biblical chronology? Should he be seen as anything more than an Arthurian-type legend? Naturally, this question is fairly important to Israelis, and it goes well beyond archaeological interest. It goes to the meaning of why the Jewish people are here in this land at all. Charles Warren's explorations are again helpful, since he noticed a cave-like chamber near the tunnel's entrance at the spring. Clearing out the chamber with help from Arab workers, he found another tunnel, which he followed some forty feet, where it ended in a peculiar shaft that rose into the darkness. While the longer tunnel was Hezekiah's, "Warren's Shaft," as it came to be called, has been the subject of much speculation through the years, giving rise to the notion that this might have been the very passageway used by young David and his men, to invade by stealth and conquer the Canaanite stronghold.

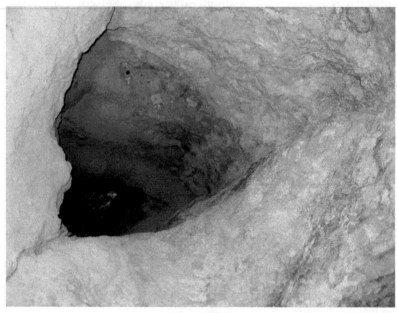

Warren's Shaft

Archaeologists have recently cast a pall of uncertainty over whether Warren's Shaft could have been used to conquer the city, as it is virtually impossible to climb. In 1995, however, as a new visitors' center at the City of David was being constructed, salvage workers underground came upon yet another secret tunnel. It led down from the ancient city proper to a water pool, and was guarded by a massive wall and towers. Bypassing Warren's Shaft entirely, this tunnel would have afforded a group of skulking invaders, David and his men, a clear line of entrance into the Jebusite city. Tiny shards of pottery littered about date this subterranean burrow to about two millennia before the Common Era, well before David lived, which means that it would have been there when the daring Israelites would have carried out their insurgency.[15] We therefore have a bona-fide archaeological link to the conquest of Jerusalem, not by the ancestors of the Palestinian Arabs, but by the distant forefathers of Israeli Jews. Yet, even the discovery of this alternate shaft by no means proves the existence of David, nor does it authenticate the city that bears

15 Hershel Shanks, "I Climbed Warren's Shaft (But Joab Never Did)," *BAR* 25/6 (1999): 31-35.

his name. Questions still linger, and they are at the heart of today's political controversy.

Digging Down, to Dig up a Palace

Is the identification of the City of David a proverbial case of "much ado about nothing," over an ancient king whose very existence is questioned? Sometimes, the simple asking of a question, as intrepid reporters are inclined to do, is enough to cast serious doubt. In 1997, third-generation Israeli archaeologist Eilat Mazar went on the hunt for any remains she could find of the actual palace of King David, which she pursued on the basis of a single verse in the Bible (2 Samuel 5:17), and which she dutifully cited in an interview she gave to the *New York Times*:

> "When the Philistines came to fight, the Bible said that David went down from his house to the fortress," she said, her eyes bright. "I wondered, down from where? Presumably from where he lived, his palace... So I said, maybe there's something here."[16]

Her reference was of course to the City of David. Mazar finally began digging in February 2005, discovering Byzantine artifacts and a mosaic floor at two meters deep. Further down were Second Temple artifacts, and beneath all of that, she came upon the foundations of a large stone structure that might possibly be the telltale remains of David's palace. Once again, Palestinian Arab detractors needed to chime in. *The New York Times* was careful to point out that Hani Nur al-Din, an archaeologist from Al Quds University, declared biblical archaeology nothing more than an Israeli attempt "to fit historical evidence into a biblical context." He went on to say, "The link between the historical evidence and the biblical narration, written much later, is largely missing."[17] Apparently, so he argued, the Bible was written "much later" than the events it records. Might there have been no "biblical Israel" at all? Are today's Israelis and the nation they founded living in this land due to what amounts to an elaborate myth? Again, one need only to ask a question to cast doubt.

Moreover, Israel Finkelstein and others joined in the skepticism,

---

16 Steven Erlanger, "King David's Palace Is Found, Archaeologist Says," *New York Times*, Aug. 5, 2005.

17 Richard A. Freund, *Digging Through the Bible: Understanding Biblical People, Places, and Controversies through Archaeology* (Lanham, MD: Rowman & Littlefield, 2009), 144.

arguing that Mazar's structure might well date to the much later Hasmonean Dynasty, of the second and first centuries B.C.E. Are the Palestinian Arab supporters in today's international media right to question the connection of the site with David? Fortunately, the facts underground will have the last word. Subsequent excavations uncovered tenth century artifacts, including imported luxury goods, within the stone structure itself. Ivory, Phoenician-style inlays attached to iron objects, a black-on-red jug dated to the tenth century, and a bone, radiocarbon dated to between 1050 and 780 B.C.E., go a long way toward establishing the Davidic "credentials" of the site.[18] Did Mazar "twist facts to suit theories" simply because she used the Bible as a "hint" on where she might dig? She has certainly taken her share of academic flak for this, especially when archaeologist Ronny Reich of Haifa University accused her of acting "unethically."[19] It should be countered, however, that if one happens to have a site that dates to a certain period, and if there is also a biblical passage that correlates with it, is one not justified in drawing certain conclusions about the veracity of the text?

The New York Times, on the heels of all the excitement, reported that progress was nonetheless impeded by the residents of the area directly above the ruins:

> Mazar continues to dig, but right now, three families are living in houses where she would most like to explore. One family is Muslim, one Christian, and one Jewish.[20]

Of course, if an organization such as the Ir David Foundation/ El'Ad actually attempts to purchase the property needed to continue the exploration, they are accused of being a "settlers'" organization, engaging in "provocative" actions. Consequently, news as earthshaking as discovering the whereabouts of King David's palace is unceremoniously subsumed under the much more "critical" issue of potential Palestinian statehood. It appears as if King David has been one-upped by modern geopolitics.

18　Eilat Mazar, *Excavations at the Summit of the City of David, Preliminary Report of Seasons 2005-2007* (Jerusalem and New York: Shoham, 2009), 52-3; H. Herzog, and L. Singer-Avitz, *Redefining the Center: The Emergence of State in Judah* (Tel Aviv: Institute of Archaeology, 2004), 209-44.

19　Ronny Reich, *Excavating the City of David* (Jerusalem: Israel Exploration Society, 2011).

20　Erlanger, op. cit.

Underground City of David excavations

Subsequently, however, a second phase of the dig was successfully launched in 2007, on a plot of land adjacent to the first. The news was significant, namely, that Mazar's "large stone structure" was even larger than first imagined, boasting walls some seven meters in thickness. Even more striking, this ancient building shows signs of being related to a stepped stone structure, famously excavated between the 1920s and 1980s. Somewhat of a landmark in East Jerusalem, it had long been suspected of being in some way connected with an edifice mentioned in the books of Kings and Chronicles, the so-called "Millo" built by King Solomon, but never described in terms of what it actually was. Archaeologists suggested that it may in fact have been constructed by the Jebusites, before David and Solomon ever arrived on the stage of history. Now, thanks to the "provocative" excavations carried out by Israeli "settlers," there is strong evidence to suggest that the Millo and Mazar's "large stone structure" belonged to a single massive royal palace. It strains credulity that the producers at CBS News seemed blissfully unaware of this angle on the excavations.

Nor did they seem aware of one more of Mazar's discoveries, in 2010, concurrent with the production of the *60 Minutes* piece. It was a wall, dated by the archaeologist to the tenth century B.C.E., and built, if not by David, then certainly by his son, King Solomon. Said Mazar,

> It's the most significant construction we have from First Temple days in Israel. It means that at that time, the 10th century, in Jerusalem there was a regime capable of carrying out such construction."[21]

Not surprisingly, detractors and debunkers immediately surfaced. Aren Maeir of Bar Ilan University was skeptical, and unsure that the remains are as old as Mazar claimed. The Palestinian Arab reaction was predictable. Nur al-Din accused Mazar of caring more about publicity than peer review:

> She doesn't give any archaeological context to her findings other than dating pottery shards. The Bible should be put aside. It's not a history book.[22]

However, is not the presence of pottery shards exactly the criterion used to establish the dating of virtually everything in the field of archaeology? Moreover, no one asked Nur al-Din whether he cared more about Palestinian Arab talking points than archaeological and historical truth. Mazar's response was pithy and powerful:

> The question is if we can trace that core [of reality] and prove it existed. Well, here it is.[23]

In the final analysis, biblical archaeology offers a powerful tool. The state of Israel is, as usual, mired in a sea of confusion, accusation, and indictment, for its very existence on what ought to be "Arab land." In such a climate, the trawl and the spade of the archaeologist provide an opportunity to "speak truth to propaganda." All that is needed is a modicum of *hutzpah*. Today, the bulk of the excavations in the City of David have been completed, but there are still plans afoot to dig in the most sensitive parts of Jerusalem, beneath the Old City and its polyglot, multiethnic quarters. There are designs to restore the true ancient city, well below the one visible to the eye. It is "Jerusalem 2.0," underground. Down there, no stones are thrown, no rubber bullets are fired. The only

---

21 Abe Selig, "Jerusalem city wall dates back to King Solomon," *Jerusalem Post*, Feb. 23, 2009.

22 Solomon, op. cit.

23 Ibid.

light is artificial, but the controversies surrounding all of this, and what El'Ad has already done to the area, are all too real.

Perhaps that is because Jerusalem is today as it has always been, a focal point of geopolitics. From the days in which David first vanquished the Jebusites, it has been attacked fifty-two times, captured and recaptured forty-four times, besieged twenty-three times, and destroyed twice. Ironically, we would be hard-pressed to find any place on earth more prone to violence and bloodshed than this "City of Peace." The fact that subterranean archaeology is on the front line in today's conflict is but a new wrinkle in the ongoing battle for David. For better or for worse, the diggers will continue to dig. Israel's greatest modern poet, the late Yehuda Amichai, summed it up:

> Jerusalem is like an Atlantis that sank into the sea
> everything there is submerged and sunken
> This is not the heavenly Jerusalem but the one down below,
> way down below. And from the sea floor they dredge up ruined walls
> and fragments of faiths, like rust-covered vessels from sunken
> prophecy ships. That's not rust, it's blood that has never dried.

CHAPTER TWO

TUNNELING HISTORY –
RABBI OF THE LOST ARK

IN SOME WAYS archaeology reminds us of the Platonic notion of the demiurge, a mysterious, artisan-like figure who fashions and maintains the material universe. In our case, archaeology brings an entire people (long forgotten) back into existence in a manner not unlike the casting of spells. Like a fire throwing shadows on the wall of Plato's cave, those who interpret the daintiest artifacts are left to ask whether the grand designs they weave are anything more than faint forms on history's elaborate tapestry. Sometimes, however, the forms magically coalesce to present us with seemingly irrefutable evidence of a people's past, stretching from time immemorial down to the present. So it is with the Jewish people and the archaeological relics that bring us back to when and where they lived in the "Holy Basin" of Jerusalem.

Such observations are brought into focus by an incessant swell of Palestinian propaganda, alleging that the central and most important edifice in all of Jewish history, the great temple originally built by the venerable King Solomon and rebuilt in the sixth century B.C.E., and rebuilt again by the most un-venerable King Herod the Great, never existed at all. There could hardly be a greater affront to Jewish sensibilities, for according to tradition, it was here that God gathered the dust from the ground to fashion the first human being, Adam, and breathed into him the breath of life. It was also in this place, known in early days as Mount Moriah, that the first of the biblical patriarchs, Abraham, is believed to have nearly sacrificed his promised son, Isaac.

Magical Mystery Tour

The sanctity of the site for Muslims is summed up by the current Grand Mufti of Jerusalem, Sheikh Muhammad Hussein, who declared: "We consider this the spot where the Prophet Muhammad began his ascent to heaven." His reference is to the Qur'an, which, in the chapter titled "Al-Isra" ("Night Journey"), relates that in a single night, when Muhammad was living in Mecca, twelve years after his divine call, a mythological winged steed known as al-Buraq took the prophet on a "magical mystery tour" to the "farthest mosque." Its location is never identified, but it was later assumed to be Jerusalem's Temple Mount, the Haram al-Sharif. Muslims today believe that the rocky outcropping on which the Dome stands is the place where, as the Quranic *sura* ("chapter") 17 continues, Muhammad dismounted his steed, prior to ascending (again on al-Buraq) into the various heavens. He received an audience with the prophets of old, and finally with Allah, who instructed the Muslims to pray fifty times a day, though relenting and reducing the number to five. Afterwards, al-Buraq returned the prophet to Mecca in the same night.

The Dome of the Rock and al-Aqsa Mosque

The story, fantastic as it is, remains central in Islamic culture to this day, though many Muslims consider it to have been merely a vision, emphasizing the purity of Muhammad's heart. Moreover, the identification of this story with Jerusalem is tenuous at best. "The Farthest Mosque" (*al-Aqsa* in Arabic) is the name of the other great structure on the Temple Mount, decorated with a blackish dome, and still used for prayer. The Dome of the Rock, by contrast, is a holy shrine, but not a proper "mosque." Critics rightly point out that the Quran's "farthest mosque" could really be anywhere, and early commentators thought of it simply as a designation of heaven. Early Muslim warriors, however, may well have reasoned that if Jerusalem is important to Jews and to Christians, there needs to be an Islamic claim on the city as well, and this may have been motive enough for linking it with Muhammad's *Isra*. In any case, the Quranic account was enough to establish Jerusalem as the third holiest site in Islam (after Mecca and Medina).

This is what the gilded Dome of the Rock, completed in the year 692 C.E. at the order of Umayyad Caliph Abd al-Malik, commemorates. Why, however, must the two previous temples which stood here be delegitimized and ultimately denied? We know that Holocaust denial has taken root in the Middle East (notably Iran), but now we have to cope with a new phenomenon, "Temple denial." The seriousness of this lunacy was revealed during the failed Camp David Summit in July 2000, hosted by President Bill Clinton. To the president's astonishment, PLO Chairman Yasser Arafat declared to Israeli Prime Minister Ehud Barak that there never was a temple in Jerusalem: "I will not allow it to be written of me that I have...confirmed the existence of the so-called Temple beneath the Mount." Arafat suggested instead that the actual site of the Jewish temple may have been in the West Bank city of Nablus, known as Shechem in biblical times. It strains both credulity and common sense to imagine such a statement proceeding from the mouth of the leader of a proposed new Palestinian state, presumably to live side-by-side with Israel in peace.

Similarly, the head of the Islamic Supreme Council for the *Waqf*, Sheikh Salad Din Alami, declared, "There are no Jewish remains on the Mount. There never were Jewish antiquities here." Sheikh Ikrema Sabri, the former mufti of Jerusalem, added:

> There was never a Jewish temple on al-Aqsa, and there is no proof that there was ever a temple. Because Allah is fair, he would not agree to make al-Aqsa if there were a temple there for others beforehand.

In 2008, Ahmen Qurei, the chief Palestinian negotiator with the state of Israel, declared:

> Israeli occupation authorities are trying to find a so-called Jewish his-

torical connection, but all these attempts will fail. The [Temple Mount] is one hundred percent Muslim.[1]

Sheikh Taysir Tamimi, Palestinian chief justice and among the most important Middle Eastern Muslim clerics, stated:

> There was no Jewish civilization in Jerusalem. Many peopled live here throughout the ages and they left some artifacts, but so what? There is no proof of any Jews being here. Jews came to the [temple area] in 1967 and not before.[2]

Presumably, in Palestinian eyes, the testimony of the ancient Jewish historian, Josephus Flavius, who witnessed the temple with his own eyes, is fictitious and holds no weight. He described it as follows:

> In the eighteenth year of his reign, Herod started to enlarge and reconstruct the temple at his own expense, which we knew would be his greatest enterprise. After removing the old foundations, he laid new ones, and raised the structure of hard, white stones. Purple hangings covered the entrances, and a golden vine with grape clusters adorned the area below the cornice. Large porticos with one hundred sixty-two Corinthian columns surrounded the temple, which was supported by walls of unparalleled size. Beyond the first court was a second, surrounded by a stone balustrade with an inscription prohibiting foreigners from entering on penalty of death.[3]

Of course questions may be raised regarding such an account, admittedly written after the temple had been destroyed by the Romans. Can Josephus' words alone be counted as trustworthy, establishing that the temple once stood on the Haram al-Sharif? Can archaeology prove that the temple indeed stood in Jerusalem? If so, should the Muslim shrine itself be viewed as an "illegal squatter" on Jewish holy ground? What about other Jewish sites in the holy city, such as David's Tomb, on Mt. Zion, declared "bogus" by the Palestinians? Does the denial of the Jewish past effectively nullify its present? As if to ensure that nothing of the Jewish temple is ever brought to light, every Israeli attempt to

---

1 See Aaron Klein, *The Late Great State of Israel: How Enemies Within and Without Threaten the Jewish Nation's Survival* (Los Angeles: WND Books, 2009), 77.

2 Ibid.

3 *Antiquities* 15:380; *War* 1:401. See Paul Maier, *Josephus, the Essential Works* (Grand Rapids: Kregel Publications, 1988), 250.

explore anywhere near the Haram al-Sharif is condemned by the Palestinians as "provocative," becoming, as it were, the "outrage *d'jour*."

Raiders of the Lost Ark

Taking the "long view," we cannot appreciate the nature of the rage without returning to the aftermath of the Six-Day War of 1967, in which Israel conquered the Sinai Peninsula, the Golan Heights, Judea and Samaria (otherwise known as the "West Bank") and, for the first time since the year 70 of the Common Era, East Jerusalem. Up until then, the area today governed by the Palestinian Authority had been annexed by the Hashemite Kingdom of Jordan. There was no move to create a Palestinian state under either King Abdullah or King Hussein, nor would such an effort have been tolerated. Israel had won independence in 1948, but her citizens had no access to the holiest sites in Judaism, since Jerusalem was a divided city and the eastern half belonged to Jordan. Suddenly, Israelis woke up to a new reality, as they listened to the recorded words of paratroop commander Motta Gur. Leading his brigade onto the sacred plateau, he had breathlessly exclaimed over a walkie-talkie: "The Temple Mount is in our hands!" The euphoria which followed was unlike anything the Jewish people had ever known during the long centuries of their worldwide dispersion. Battle-hardened soldiers were among the throng of Israelis who now streamed to the Western Wall to pray. Unable to contain themselves, they could only stand before the massive stones, weeping.

Not surprisingly, a group of Orthodox Israeli Jews, led by Western Wall "rabbi emeritus" Yehuda Getz, began to nurture a secret desire to get as close to the ancient shrine as possible. Getz was a complex individual, a secret philanthropist of deep spirituality, as well as an officer in the Israel Defense Force. His looks matched his mystical approach to his many endeavors. Always clothed in black with a white head scarf, he carried a Bible and a prayer book in his pocket, while armed with a pistol in a side holster. When the Jewish quarter of the Old City was conquered by Israel in 1967, he was one of the first Israelis to resettle there. Suddenly, the idea of excavating along the Wall was no longer an unrealizable fantasy.

The area to the immediate left, while facing the Wall, was covered by centuries of debris. Above the landfill, the Muslim Quarter of Jerusalem had taken shape, dating back to the seventh century, when Arab armies had taken the city from the Byzantines. In 1967, Rabbi Getz (who coopted a dedicated corps of young Yeshiva students) was authorized by Israel's Ministry of Religious Affairs to begin digging a fresh tunnel, horizontally and northward, along the previously covered section of the Western Wall. Archaeologists at the time were uninterested, and governmental oversight was lacking, but the digging continued, unabated, for decades. In 1981, the tunneling rabbi chanced upon a long-sealed ancient entrance to the Temple Mount, known as Warren's Gate, situated about forty-six meters from the tunnel's entrance. Originally identified by the illustrious Charles Warren, its outlines were clearly visible among the massive Herodian ashlars. In due course, Getz converted the area opposite Warren's Gate into a small synagogue, which can still be seen today. Known as "the Cave," Rabbi Getz would pray there each morning, believing it to be the point closest to the location of the temple's most sacred inner sanctum, the Holy of Holies.

Western Wall Tunnel - Secret Passage

The revered rabbi and his disciples realized that they had come upon an entrance to an enormous complex of subterranean vaults that honeycomb the artificial portion of the great plateau, upon which Herod had erected his vastly refurbished Jewish temple. They clearly believed that the cavernous underground chambers on the other side might even lead them to the foundations of the temple (theorized to lay directly beneath the Muslim shrine known as the Dome of the Rock), including the Holy of Holies. Rabbi Getz and his team now turned their digging equipment on the gate itself. They managed to break through the subterranean portion of the ancient rampart, finding an enormous carved tunnel, six meters wide and twenty-eight meters long, leading eastward toward the remains of the temple foundations. The floor was covered with mud and water.[4] Getz wrote in his journal:

> I immediately ran to the site and was overwhelmed with excitement. I sat there for a long time sprawled helplessly as warm tears were streaming down my face.[5]

Looking past the virulent anti-Jewish Palestinian screed of "temple denial," we are told in the Hebrew Scriptures that the first "Holy House," long pre-dating Herod, was commissioned by the wise King Solomon, around 1,000 B.C.E. The fabled Ark of the Covenant was said to have once resided within, nestled securely in the most holy chamber, having been brought up to Jerusalem by none other than Solomon's father, the mighty King David. However, forasmuch as the golden chest was celebrated, it inexplicably vanished from the biblical narrative, never to be referred to again, except in non-canonical works such as 2 Maccabees. In that apocryphal text, it was said to have been spirited out of the city by the prophet Jeremiah, just before the temple's destruction by the Babylonians, and hidden away in a desert cave until the end of days. Other legends attempt to trace its melancholy journey to a church in Ethiopia, while a number of "pop scholars," crawling onto the fragile limb of speculation, postulate that Judaism's most sacred relic was lowered into a subterranean cavity during the siege of the city, escaping the fiery conflagration which engulfed the sanctuary in 586 B.C.E. If that were the case, then the most important holy relic in world history might still rest in the vaulted chambers beneath the Temple Mount platform. Indeed, Rabbi Getz claimed that during this forbidden foray, he actually laid eyes on the hallowed Ark but that he covered its position and never discussed it again, so as not to be targeted by the Muslims.[6]

Meanwhile, atop the expansive plateau of the Haram al-Sharif, the

---

4 See *JPRS Report: Near East & South Asia,* Issue 91052 (Foreign Broadcast Information Service, 1991), p. 72.

5 See Gershom Gorenberg, *The End of Days: Fundamentalism and the Struggle for the Temple Mount* (Oxford: Oxford Univ. Press, 2000), 125. https://cryforzion.com/has-the-ark-of-the-covenant-ever-been-found/.

6 Ibid.

curious sounds of digging, rising up through one of the many cisterns in the vicinity of the Dome of the Rock, were overheard by Palestinian workmen, who reported the underground activity to guards from the *Waqf*. A number of stalwart young men were immediately dispatched through the entrances to the cisterns, carrying cinderblocks and trowels to erect their own wall for protection from these Jewish "invaders." They came face-to-face with Rabbi Getz and his team, engaged in clearing the debris from the gate complex and the adjacent vaulted passageway veering off into the darkness. In an attempt to "discourage" the rabbi's adventure, physical violence erupted, and the two sides came to blows. There were multiple injuries, while up above, the *Waqf* incited rioting across Arab East Jerusalem.

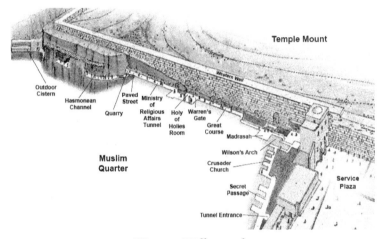

Western Wall tunnels

Why would Islamic authorities (the *Waqf*) consider the rabbis' search for secret treasures, including perhaps the Ark of the Covenant, an unthinkable provocation, for which calls for holy *jihad* are issued? The grounds for incitement have included the idea that such subterranean digging might endanger the foundations of the golden dome and other Muslim structures, including the Al-Aqsa Mosque. Another charge is that Zionist intruders might even attempt the unthinkable – detonating explosives underground so as to bring down the holy sites. To be sure, such fears are not entirely unfounded, since, as we shall see, certain fringe elements in Israel, spurred by religious fanaticism, have in fact plotted to bring down the Dome of the Rock. But beneath the motivation to protect sacred shrines, a desire shared by Israeli authorities with regard to all religious sites in the Holy Basin, lies a deep-seated Palestinian rejection of the Jewish state itself. A cursory look at history tells us that while Jews were to be tolerated in Islam, they were to be in a special category, designated as *ahl ul-dimmah* ("protected people"), whose "protection" meant living forever as second-class citizens.

As observed by orientalist scholar David Farhi, the toleration of Jews in the Muslim world was contingent upon their being an enslaved people, whose rights did not extend to the political arena.[7] Another oriental scholar, Moshe Sharon, pointed out that the modern Jewish nation was created in the heart of the so-called "Dar al-Islam," namely, land on which holy sites are located. It is worth noting that there is no separation of religion and state in the Islamic world. The prophet Muhammad declared, "Religion and the state are twins." Jews are to be beneath and subordinate to Muslims, never being allowed to rule over them. As recounted in countless Friday sermons delivered in mosques for decades, Jerusalem is to be a Muslim city, the third holiest in the Islamic world.

It takes little imagination to understand why any form of archaeological exploration at or near the Haram al-Sharif would be considered as undermining not just the foundations of the structures above, but the entire fantasy of Muslim sovereignty over the entire city. The creation of the *Waqf* under Israeli Defense Minister Moshe Dayan was a clear and some say misguided attempt to placate Muslim sensibilities, for the bulk of the Palestinian population continues to reject the sovereignty of the Jewish people in their ancient land.[8] Any attempt, archaeological or otherwise, to demonstrate the historical Jewish presence in this land must be met with the fiercest resistance and defeated at all cost. In the end, the Israeli government ordered the sealing of the gate opened by the tunneling rabbis with six feet of reinforced concrete. Would this be enough to satisfy the *Waqf*? Hardly, especially since Rabbi's Tunnel, now known as the Western Wall Tunnel, was extended, ever northward during the years to come.

Tunnel of Terror

More archaeologically-linked unrest was destined to erupt. As the din of indignant Palestinian voices slowly subsided, Israel's Ministry of Religious Affairs cautiously allowed the northward digging of

---

7 See David Farhi, "The Muslim Council in East Jerusalem and Judea and Samaria Since the Six-Day War," *ha-Mizrah he-Hadash*, 28/1-2 (1979): 3-21.

8 See Yitzhak Reiter, *Contested Holy Places in Israel–Palestine: Sharing and Conflict Resolution* (London: Routledge, 2017).

the Western Wall Tunnel to proceed, unabated. To ensure no damage to the structures above, safety engineers were engaged. Along the way, this new round of digging happened upon the most massive Herodian ashlar ever discovered on the Temple Mount, or anywhere else in the land of Israel. It is a single stone block, 13.6 meters in length, between 3.5 meters and 4.5 meters in width. It is estimated to weigh a staggering five hundred seventy tons, and is among the heaviest objects ever raised by human beings without mechanized cranes. Proceeding further along the newly revealed courses of finely chiseled limestone, the persistent diggers eventually arrived, years later, at the northern extremity of the vast Temple Mount platform.

Western Wall Tunnel with largest Herodian ashlar

At this point the artificial tunnel being excavated suddenly intersected with an even more ancient water channel, dating to the pre-Herodian Hasmonean Dynasty, of the first two centuries before the Common Era. Significant amounts of water would in fact have been necessary in those days, in order to cleanse and purify the area around the temple's altar, soaked on a daily basis with the blood of sacrificial animals. The period of this "Second Temple" was the last historical moment that the Jewish people were truly free in their own land (save for two catastrophically failed anti-Roman revolts), until the birth of the State of Israel in 1948. Indeed, today's Israelis live directly on top of such reminders from antiquity, with that which lies beneath constantly fueling the desire to resettle the whole of their ancient homeland – to create "facts on the ground." Faced with the true stories in stone which dot this Holy Land, the Palestinian Arab claim that they are its true natives, displaced and disenfranchised by white European colonists who happen to be Jewish, is on its face patently absurd. Indeed, a majority of today's Israeli Jews are partly descended from Middle Eastern or North African families.

In any case, while the Western Wall Tunnel and the so-called Hasmonean Tunnel were stabilized and officially opened to the public in 1988, there was no exit, and those who made the long trek through its dimly lit recesses were compelled to retrace their steps, some three hundred eighty meters, back to the beginning. Significant interest began to develop around the possibility of opening a northern exit from the tunnel. Israeli authorities approached the Sisters of Zion convent on the famed Via Dolarosa (the traditional path on which Jesus carried his cross to Mount Calvary), in the hope of finding a spot for the diggers to break through. The Sisters, however, declined, not wanting to become involved in the geopolitical ramifications of a simple archaeological dig.

Next, the Arab owner of a souvenir shop on the Via Dolarosa was approached. Surprisingly, he agreed, given that crowds of tourists exiting the tunnel at his shop could hardly be bad for business. Nonetheless, after some consideration, he too declined, since his fear of being targeted by his Palestinian compatriots trumped his profit motive. To be sure, the murder of Palestinian "collaborators" with Israel – by fellow Palestinians – goes largely unnoticed, even while UNESCO busies itself with its next official condemnation of the Jewish state. Israeli authorities nonetheless persisted in their attempt to find an exit for the tunnel, next considering an opening on the street itself. The issue was elevated directly to the office of the prime minister at the time, Yitzhak Shamir, who vetoed the plan, due to concern over the unrest that might well erupt.

In 1996, however, another Israeli prime minister, Benjamin Netanyahu, decided that the long years of delay must come to an end. Possessing considerably more archaeological chutzpah than his predecessors, he authorized the construction of a staircase leading up from the tunnel to a spot on the Via Dolorosa adjacent to a school. Officials of the *Waqf* were given a tour of the tunnels and an opportunity to examine maps, indicating their precise location and proving that they in no way undermined the Haram al-Sharif. They were also invited to open a new gate to Solomon's Stables (an underground vaulted area at the bottom of a set of stairs descending from the al-Aqsa Mosque, beneath the Temple Mount), as well as the right to hold Muslim religious services there.

None of this mattered, as Jerusalem mayor Ehud Olmert, sledgehammer in hand, joined in breaking through to the street above.[9] The tunnel's new exit was separated from the street itself by an ordinary-looking stone wall, but the inconspicuous opening from below did nothing to mitigate the fierce reaction from Jerusalem's Arab inhabitants. Moreover, the *Waqf* was alarmed by the prospect of fresh hordes of Jewish tourists being belched forth from the depths of the earth into the heart of Jerusalem's Muslim quarter.

It is truly odd that UNESCO never deigned to condemn the anti-Semitic attitudes of the Palestinians, outraged as they were that Jews might dare to surface from beneath the ground on land which belonged to the ancient Israelites long before Islam ever existed. No one should need reminding that in the United Nations, double standards rule. In the Security Council, following a complaint by the representative from Saudi Arabia that Israel had opened a tunnel "in the vicinity of the al-Aqsa Mosque," Resolution 1093 was adopted, once more condemning the Jewish state. Of course the tunnel has nothing to do with the mosque, nor did it violate Israel's Interim Agreement with the Palestinians, which does not cover archaeological projects. At the U.N., however, facts hardly matter, and anti-Israel resolutions are piled from floor to ceiling. If the tunnel excavation proved anything, it was to underscore that the Muslim Quarter had obviously been built directly on top of land seized by conquest from the Jewish people long ago. It was the Arabs who were relative latecomers to Jerusalem, not the Jews.

The tunnel was opened on a Tuesday, prompting immediate rioting in Jerusalem and other locations. It quickly turned deadly, as Israeli police employed increasingly forceful measures to defend themselves. Amid a growing unrest, Netanyahu called in the I.D.F., deploying tanks and attack helicopters. By Thursday, at least sixty-eight people had died, mostly Palestinian, though several Israelis were also shot dead. Of course, it was Israel that received growing international condemnation. Prime Minister Netanyahu summed up his government's position:

> Unfortunately, what we saw here was an attempt to cynically manipulate a non-issue: a fabrication that says that we in any way affected or hurt the Islamic holy places... The underground tunnel does not do that in any way. The chairman of the Palestinian Authority and his spokesmen knew exactly that. We always respected the Muslim holy places.[10]

---

9   See Gorenberg, op. cit., 182 ff.

10   Storer H. Rowley, "Netanyahu: Riots Aren't Israel's Fault," *Chicago Tribune*, Sep. 28,

By the end of the week, however, more trouble was brewing. In anticipation of what was to come, a squad of Israeli security forces, having donned riot gear, assembled on the Haram al-Sharif. Simultaneously, a mass of some ten thousand Arab rioters, having finished their Friday prayers, began to vent their rage, hurling stones at the police and, worse still, at Jewish worshipers down below, praying at the Western Wall. As if on cue, the security police charged into action, firing tear gas, followed by rubber bullets, at the advancing rioters. Fifty were wounded, including five Israel policeman, in the ensuing ruckus. The Palestinian account amounted to an exercise of orchestrated incitement. While the Israelis on the scene reported "massive stone throwing," the other side declared that only a few stones had been hurled and that the Israelis had responded with live ammunition, killing three of the protesters.

PLO Chairman Yasser Arafat, who many argue was himself an arch terrorist with Jewish blood on his hands, declared "… the attempt to attack worshipers inside the mosque is something that cannot be tolerated."[11] Such words certainly had an effect, even though Arafat later said that he had ordered his underlings to prevent attacks on Israelis. The incitement, however, ran its logical course, with extensive rioting erupting in other Arab neighborhoods as well as other cities in the Palestinian territories. In the vicinity of Jericho, three Palestinians were killed during the unrest, and in Tulkarm, a Palestinian and two Israeli paramilitary officers were slain. An Israeli colonel was killed in Rafah, near Gaza's border with Egypt, and on the other side of the border, an Egyptian police Lieutenant was killed by stray fire from an Israeli helicopter gunship. Israel dispatched Cobra helicopters to Ramallah, while tanks took up position opposite Nablus. Many more Palestinians were wounded in incidents in Hebron, Qalqiliyah, and Bethlehem. To be sure, the ongoing Arab propaganda war was not about to subside. In Gaza, Arafat thundered: "Our blood is cheap in the face of the issue for which we are gathered here."[12] On Palestinian radio, a caller announced the need "… to slaughter all the Jews [and] to appoint a caliph for Palestine."[13] Incredibly, all of this was over nothing more than a tunnel and a new access route to an archaeological treasure.

1996.

11 Serge Schmemann, "10 More Die in Mideast as Violence Erupts 3d Day; Mosque is Scene of a Clash," *New York Times*, Sep. 28, 1996.

12 Nadav Shragai, "The Seeds of Calamity," *Ha-aretz*, Sep. 27, 1996.

13 Ibid.

Ancient Sewer, Modern Muck

Recently, as another tunnel, this one connected to one of the most tragic events in Jewish history, was being prepared for a "grand opening," an exiled leader of the terror group Hamas declared that Israel was "playing with fire" – sadly, "par for the course" in today's Middle East, as archaeology and politics butt heads. The modern controversy, like so much else in the ironically dubbed "city of peace," is rooted in antiquity's mists, yet precipitated only recently, beginning with an accidental discovery. The year was 2004. Location: the "Holy Basin," a compact area of just one square kilometer, where some of the most important sites of Judaism, Christianity and Islam are located. At the southern extremity of the ancient City of David, a subterranean sewage pipe had burst. The municipality of Jerusalem promptly dispatched a work team to make appropriate repairs. Due to the sensitive nature of the locale, a small cadre of competent archaeologists accompanied the workers. As work progressed and the ground was penetrated, the digging was punctuated by the sound of scraping - steel against stone.

Work immediately halted, and another kind of excavation, archaeological in nature, commenced in earnest. In due course the remnants of a stair-lined road of sorts, clearly ancient, began to emerge. These, however, were no ordinary steps. The archaeologists were aware of another set of steps, nearly identical, leading up to the southern end of the Haram al-Sharif, the "alleged" Temple Mount, and dated to the time of King Herod the Great, in the first century before the Common Era. Nowhere else in the land of Israel have steps of this particular style and character been discovered. Might this "road" of steps have been directly connected in some way with the Jewish temple that once stood where the Dome of the Rock now stands? It appeared, moreover, that these newly excavated steps had, in bygone days, led down to the ancient Pool of Siloam, revered in Christian Scripture as the place where Jesus of Nazareth commanded a man "blind from birth" to wash, thereby returning to him his sight. There was indeed good reason for what was essentially an enormous ritual immersion bath (*mikveh*), as large as two Olympic-sized swimming pools, to have been situated in precisely this location, since pilgrims ascending to the holy hill were first required to attain ritual purity. According to estimates, well over two million people (certainly including Jesus of Nazareth) would, during Jewish festivals, have made their way up to the temple (non-existent, according to Palestinian screed) subsequent to immersion in the pool.[14]

Remains of Second Temple Pool of Siloam

The destination, which was situated some considerable distance to the north, was an enormous artificial platform supported by multiple rows of subterranean vaults, which effectively extended the original hill identified in Solomon's day as Mount Moriah. This "Court of the Gentiles," as it was called, covered an area of some 480 x 300 meters, one side of which includes the Western Wall and the adjacent underground tunnel, which had been a flashpoint of controversy and violence in years past. The southern end of the temple complex has, since the conquest and reunification of Jerusalem in the 1967 Six-Day War, been thoroughly excavated. Rows of shops, still visible, were constructed along the base of the Temple platform on several sides, along with the remains of the great "Double Gates" and the "Triple Gates" that once served as the entrance (via underground ramps) to the great courtyard above.

In order to make the seven hundred meter trek from the Pool of Siloam to the upper temple precincts, throngs of Jewish worshippers would have traversed the stair-lined road of pilgrimage, beneath which was a subterranean drainage tunnel, now entirely revealed to the amazed

---

14 Zëev Orenstein, Director of International Affairs for the City of David, interview on *Mark Levin Podcast*, July 1, 2019.

archaeologists.[15] An estimated ten thousand tons of quarried rock had been required in the construction of the road itself, which, according to archaeologist Nachson Szanton, could be safely dated to a time no earlier than 30–31 of the Common Era, concurrent with the Roman prefect Pontius Pilate.[16] Touring the site, Szanton observed:

> Every step on the street brought the pilgrims closer to the temple. Imagine … the joy, the songs, the prayers, the spiritual journey that these people experience when they know they are just meters away from reaching the gates of the temple.

Following clues provided by archaeological maps and charts spanning the last century and a half, the digging proceeded apace, in a manner that would have dazzled Charles Warren himself. Dozens of fiber-optic camera cables were utilized in the effort to determine the proper places for excavation, while engineers were brought in to stabilize the tunnel and to ensure that none of the structures above would sustain damage. Along the way a stepped structure was uncovered, likely put to service as an ancient podium (*bemah*), from which religious and political figures would hold forth in oratory.

As the ground was cleared, what was found confirmed the tragically melancholy tale, told by Josephus Flavius, of the great Jewish Revolt against Roman domination of the land of Israel. In the year 66 of the Common Era, the notorious Nero Caesar, ensconced in the city on the Tiber, dispatched his most illustrious General Vespasian to put it down. In the year 70, the glorious temple of King Herod was burned to the ground. Josephus wrote: "Around the altar were heaps of corpses, while streams of blood flowed down the steps of the sanctuary." A mass of Jewish refugees fled below ground, to the drainage tunnel, beneath the pilgrimage road, now being excavated. The Roman legionnaires, discovering their whereabouts, broke through the road with massive hammers, entering the subterranean sewer, and unleashing carnage on some two thousand hapless refugees. So ended the siege of Jerusalem, which was more than the suppression of an uprising. It was a holocaust, said by Josephus to have resulted in more than one million Jewish casualties.

---

15 See Amanda Borschel-Dan, "On an Ancient Road to the Temple, Archaeological Innovation, Mystery and Dispute," *The Times of Israel*, Oct. 17, 2019.

16 See Owen Jarus, "Archaeologists Identify 'Lost' Jerusalem Street Built by Pontius Pilate," *Live Science*, Oct. 21. 2019: https://www.livescience.com/pontius-pilate-street-jerusalem-found.html.

Model of ancient Jerusalem showing drainage channel

What the modern team of archaeological sleuths discovered uncannily confirmed the description of the ancient historian. Cooking pots and other domestic items littered the tunnel's dark recesses, suggesting that the refugees had been sheltering there for days, weeks, or even months prior to their merciless slaughter, as attested by the presence of a Roman sword and scabbard, abandoned in *situ*. Most hauntingly, the huddled Jewish inhabitants of this last hiding place in the city scratched into the wall an etching of a menorah, the golden lamp stand which once adorned the temple's inner chamber. Greeting the eyes of the modern team for the first time in two millennia, they instantly recognized that this was the earliest depiction of the sacred menorah ever to come to light.[17] There were also hundreds of coins, struck with the Hebrew words: "For the freedom of Zion." It was as if those refugees, from their shrouded gloom, had telegraphed a declaration, across twenty centuries, that there was indeed a great Israelite temple here, and that this land, notwithstanding conquest and pillage by countless invading foes, is the one place on earth that belongs to the Jewish people.

Nonetheless, the ongoing excavations were to be dogged by controversy, for they were located in an area of Arab East Jerusalem, officially over the Green Line, that marked the demarcation of the city into Jewish and Palestinian sectors prior to Israel's annexation. As expected, the cabal of anti-Semitic ambassadors at the United Nations were no more impressed by messages from the past than by the reasoned arguments of modern Israeli archaeologists, laboring tirelessly to uncover the tunnel and its secrets. UNESCO brazenly condemned the new archaeological activity taking place in the Holy Basin. This, after all, is supposed to be Arab land, part and parcel of East Jerusalem – the capital of the future Palestinian state.

Opening Day

After decades of Israel-bashing, advancing an uninterrupted narrative of dark and olive skin minorities being cruelly oppressed by Western colonialists, the United States unceremoniously withdrew its participation and membership in UNESCO. Moreover, when the tunnel was ceremonially opened to the public, in June 2019, the Trump Administration dispatched its ambassadors to France, Denmark and Portugal, along with presidential adviser Jason Greenblatt and the U.S. Ambassador to Israel, David Friedman. Among the crowd of nearly one hundred was Sara Netanyahu, Israel's ambassador to the United States Ron Dermer, former mayor of Jerusalem and Knesset (Parliament) member Nir Barkat, Senator Lindsey Graham, and Republican Party donors Miriam and Sheldon Adelson.

A special papier-mâché wall was constructed for the ceremony, to be broken through by Ambassador Friedman, wielding a diminutive sledgehammer. It was a commingled moment of pathos and celebration, commemorating the sacrifice of so many lives on the spot, while adumbrating the tenacity of the Jewish people, who yet live. Indicating tacit American recognition of Israeli sovereignty over East Jerusalem, Friedman remarked:

---

17 https://www.cityofdavid.org.il/en/archeology/finds/rare-menorah-etching.

It confirms with evidence, with science, with archaeological studies that many of us already knew, certainly in our heart – the centrality of Jerusalem to the Jewish people. The spiritual underpinnings of our society, the bedrock of our principles in which we honor the dignity of every human life came from Jerusalem. This place is as much a heritage of the U.S. as it is a heritage of Israel.[18]

Inauguration of Pilgrimage Road in the City of David, June 2019.

Following Friedman's address, Barkat stood up and declared that this archaeological excavation will "… hopefully [allow] the world [to] understand why we will never, never divide the city of Jerusalem." To be sure, in today's Israel, it might be observed that archaeology and Zionism are "joined at the hip."

Not surprisingly, more vociferous condemnations followed swiftly. The foreign ministry office of the Palestinian Authority accused Israel of "imperialistic Judaization plans" designed to change the "status quo" in the holy city. The chief negotiator of the Palestinian Authority, Saeb Erekat, tweeted:

> I hope all the world, including Americans, can see this. That is not a U.S. ambassador; that is an extremist Israeli settler, with Greenblatt also there, digging underneath Silwan, a Palestinian town.[19]

The Donald Trump administration was accused of "fully supporting the

---

18 Jacob Magid, "Under Palestinian Homes, US Envoys Hammer Open an Ancient East Jerusalem Road," *The Times of Israel*, Jun. 30, 2019.
19 Ibid.

imperialistic settlement enterprise led by the far-right in the occupation state."

Israel of course is a democracy, the only democracy in the Middle East, the result of which being that voices of dissent among Israelis, Including the secular left, who have no interest in claims arising from the land's archaeological heritage, are given free expression. The leftist group, Peace Now, which favors unconditional Israeli withdrawal from the territories conquered during the Six-Day War of 1967, opined that the opening of the tunnel was effectively turning the Arab neighborhood of Silwan "... into the messianic Disneyland of the far-right in Israel and the United States – several meters from the al-Aqsa Mosque and the temple mount."[20] They dubbed the Pilgrim's Road the "Controversy Tunnel," declaring that it had been "...dug under the homes of Silwan residents, caused the evacuation of Palestinians' homes in the neighborhood and increased the tensions between Palestinian residents and the Jewish settlers acting more intensively than ever in recent years to Judaize the neighborhood, as part of an effort to sabotage the two-state solution."[21]

Another left-wing Israeli organization, Emek Shaveh, intent on delegitimizing the use of archaeology to advance Zionist goals, broadcasted its own criticism of the project: "...the horizontal excavation method, and the paucity of scientific publication, do not allow us to know for sure when the street was built and how it was integrated into the urban layout of Jerusalem." Greenblatt's response to all of this brilliantly epitomizes the nature of Jewish settlement of the ancient Jewish homeland:

We can't "Judaize" what history/archaeology show. We can acknowledge it and you can stop pretending it isn't true! Peace can only be built on truth.[22]

Against this backdrop, a *New York Times* correspondent tweeted his own condemnation, repeating the propaganda points that Palestinian homes above were being undermined and damaged by the excavations beneath. Not the slightest concession was made to the overwhelming

20 https://peacenow.org.il/en/the-disputed-tunnel-in-silwan-inaugurated-with-american-support.

21 See Jacob Magid, "Palestinians Slam US 'War Crimes' after Envoys Open East Jerusalem Tunnel," *The Times of Israel*, Jul. 1, 2019.

22 Tal Polon, "Greenblatt: We Can't 'Judaize' What History Shows," *Arutz Sheva*, Jun. 30, 2019.

care taken by archaeologists and engineers alike to maintain the structural integrity of modern buildings and infrastructure above while allowing access to ancient Jerusalem below. Ambassador Friedman quipped that we seem to know how to build subways without damaging anything at ground level; uncovering an ancient tunnel ought not to be an impossible task. Nonetheless, the same *New York Times* correspondent, in a longer article, described the "sledgehammer" used by Ambassador Friedman to symbolically break into the tunnel as evocative of the destruction of the "peace process" with the Palestinians.

Such a peace process would of course imply the division of the holy city into Jewish and Arab sectors, based on the 1948 armistice. It would place an international border through the heart of Jerusalem, effectively turning it into the Berlin of the Middle East. It would require that Israel surrender sovereignty over the holiest site in Judaism, the Western Wall, the last vestige of the Jewish temple which Palestinians claim never existed. Nonetheless, the fact of the matter is that almost every day archaeologists are uncovering the serious connections between the Jewish people and the city of Jerusalem, across millennia, contradicting the narrative that modern Zionists, representing European interests, essentially "stole" Palestine from its Arab inhabitants. As we have seen, however, the very name "Palestine" was invented by Rome's brutal conquerors, who lay waste the land and murdered her inhabitants, almost to extinction. That the Jewish people survive at all is one of history's greatest anomalies. They are, to be sure, "back home," and they are not going away...

CHAPTER THREE

BULLDOZING HISTORY –
DIGGING UP TROUBLE BENEATH THE
MOSQUE

A T THE CORE of archaeology is an inscrutable narrative, related by fractured shards of pottery, deposited in the layered debris of millennia. As in paleontology, where dinosaur bones scattered among eons of geological strata relate the ebb and flow of life's ever-changing design, the shards h ave their own story to tell. As with the fossilized bones, however, discerning the truth is not an exact science, but is the product of its interpreters. Each time another shattered fragment of some ancient household utensil emerges from the desiccated dust of an archaeological site, innumerable questions dance through the mind of the spade-wielding excavator. Whose hands touched it last? What was their world like? Who were their children, their friends and neighbors, their enemies? What terrors confronted them? What impenetrable destiny awaited them? What messages might they convey to us, across the shifting centuries, of conquest, exile, and rebirth in a modern state called Israel?

In the last decade of the nineteenth century, eccentric British Egyptologist Sir William Matthew Flinders Petrie declared that the reconstruction of the ancient past is best accomplished not by considering gigantic monuments left behind, but by piecing together the tiny remnants of broken earthenware, the "unconsidered trifles." Over time, he earned a telling moniker – "the father of pots." Yet, in the last decade of the nineteenth century Petrie was destined to stumble upon something that, while previously unconsidered, was hardly a trifle. It would in fact write a new chapter in the contentious field of biblical archaeology. Having made his way to Egypt to begin excavating the fabled city of Luxor, he discovered, in the funerary temple of the great pharaoh Merneptah, an inscribed stone. Covered in ancient Egyptian hieroglyphic characters, the black granite stone slab called a stele stands over seven feet tall. It celebrates the pharaoh's triumphant campaigns of conquest. The gods Mut and Horus appear prominently, and two images of the god Amon face outward to the pharaoh. Toward the end of the inscription, in the twenty-fourth line, we find a single staggering proclamation: "Israel lies desolate; its seed is no more." Dating to the thirteenth century B.C.E., the stele represents the oldest mention of the word "Israel" anywhere in the world.

The Merneptah Stele

Not surprisingly, a number of "minimalist" scholars and archaeologists have cast doubt on the significance of the stele and even on whether it has been properly read. A great deal is at stake, certainly for modern Israelis, who firmly believe that they have inherited a land handed down from their ancestors, the Patriarchs. Others, including those on the Palestinian side of the modern conflict, beg to differ. Can we really know who lived in this Holy Land so long ago? Was there really an ancient people known collectively as Israel, who departed from Egypt under a mighty deliverer known as Moses, and who subsequently conquered the land of the pagan Canaanites, making it their own? Was there really a venerable king named David, who established Jerusalem as the eternal capital of the Jewish people? Once asked, such questions seem to multiply exponentially. When it comes to resolving them, there is no place more archaeologically significant, more laden with "unconsidered trifles," than Jerusalem's Temple Mount.

The Southern Wall: Recalled to Life

Until only the last few decades, the southern end of the gargantuan limestone platform on which the gold-encrusted Israelite shrine once stood lay in long-forgotten repose, buried under the accumulated refuse of nineteen brutally harsh centuries. The top courses of the massive Herodian wall were visible to passersby, but what lay beneath the broad rock-strewn turf stretching out to its immediate south was a seemingly arcane mystery. From the birth of biblical archaeology in the nineteenth century, the entire area was strictly off-limits to the spade and the trowel, being perilously close to the al-Aqsa Mosque, which is perched high above on the vaulted Herodian extension of the Temple Mount (the Haram al-Sharif). "Allah forbid!" that anyone interested in Jewish remains might begin an excavation in the valley below Islam's third holiest site. What might they be trying to uncover – evidence of a Jewish temple which (so they claimed) never existed? Under Ottoman control, the very idea of digging anywhere in proximity to the Haram was unthinkable. When Ottoman rule was eclipsed by the British Mandate for Palestine in 1917, Muslim sensibilities were respected, especially since East Jerusalem is ethnically Arab, and stoking religious unrest in the holy city was the last thing His Majesty's Government needed.

During Israel's War of Independence in 1948, West Jerusalem was tenaciously held by its Jewish inhabitants, but East Jerusalem and the entire Old City fell to the Arabs, and was subsequently annexed by the Hashemite kingdom of Jordan. The historic Jewish Quarter was ransacked and demolished, down to its foundations. The centuries' old Hurvah Synagogue, a looming landmark from the Middle Ages, was dynamited and reduced to rubble. By contrast, Islam's sacred shrines (the Dome of the Rock and the Al Aqsa Mosque), were not only held in highest reverence, but no archaeology of any kind was to be permitted in their vicinity. In fact serious archaeology ceased across the whole of Judea and Samaria, the so-called West Bank of the Jordan River.

Days after Israel's unification of Jerusalem, 1967

The year 1967 had the potential to change all of that, as Israel took control of the entire city for the first time in two thousand years. Israeli authority might have immediately been extended over the Temple Mount, the government brushing aside Arab complaints as contemptuously as the Arabs had desecrated the Jewish Quarter and the Jewish cemetery on the Mount of Olives, gravestones of which had been uprooted and reused as latrine floors, urinal walls and pavement stones.[1] The whole of East Jerusalem might have been opened to excavation, but Israel chose instead to perpetuate a fragile archaeological status quo. Though wars, like elections, have consequences, the *Waqf* was left to operate as a "Vatican state" of sorts on the Temple Mount, the Israelis themselves relinquishing authority over the holy hill and intervening only during violent disturbances.

Nonetheless, in February 1968 two prominent Israeli archaeologists, Benjamin Mazar and Meir Ben-Dov, commenced a major excavation at the foot of the southern wall, extending around to the southernmost

---

1 https://harhazeisim.org/shameful-dereliction/.

corner of the hallowed Western Wall.[2] The digging went on for nearly a decade, until 1977. Surprisingly, the first opposition was voiced, not by the Arabs, but by Israeli authorities the moment the dig began. The Ministry of Religious Affairs, the selfsame body responsible for the Western Wall Tunnel, immediately objected, claiming that the entire area was a place of prayer. The Israel Antiquities Authority, however, had some leverage of its own, since every excavation was required by law to be accompanied by qualified government-licensed archaeologists. When it came to Rabbi Getz and crew, this technicality had been conveniently overlooked. Mazar and Ben-Dov, however, were certainly qualified, and it was agreed that they would be allowed to continue their dig as long as no objection was raised concerning the tunnel.[3]

Temple Mount Southern Wall

Predictably, there arose voices of scathing censure from Jerusalem's Arab population. The *Waqf* issued a vituperative series of complaints to UNESCO. It was alleged that the work was directly endangering the foundations of the al-Aqsa Mosque, potentially bringing about a collapse of the entire retaining wall and its complex vaulted support system. Nevermind that it was not only Jewish remains that would potentially be uncovered at the site, but remnants of the city's Muslim heritage

2 See Eilat Mazar, *The Complete Guide to the Temple Mount Excavations* (Jerusalem: Shahom Academic Research & Publication, 2002).

3 http://www.templemount.org/tunnel.html.

as well. To be sure, if the *Waqf* had had its way, multiple centuries of the legacy of Islam in Jerusalem would have remained buried under several meters of debris.

In 1969 UNESCO's representative, Dr. H. J. Reinink, visited the site and reported back that the dig did not threaten the structural integrity of any structures and that the excavations were of great significance for appreciating the long and complex history of the city. The subsequent meeting Reinink facilitated between the head of the *Waqf* and Mazar helped to ease simmering tensions among the Arabs, which should never have been inflamed in the first place.[4] Moreover, it should have been evident from the outset that such excavations, carried on by distinguished archaeologists and engineers, endangered neither wall nor mosque, only perhaps the nationalistic pride of anti-Semitic, "temple-denying" Palestinians.

Ironically, the most important discoveries from the early phase of the excavations were Islamic. As the archaeologists meticulously cleared the dust, dirt, rock and rubble from the area adjacent to both the southern and southwest quadrant of the Temple Mount, several magnificent structures of monumental scale, sandwiched between elaborately paved streets, began to be revealed. They were remnants of the mighty Umayyad Dynasty, which ruled the region in the seventh and eighth centuries and which established a royal quarter in this part of Jerusalem.[5] One of the most imposing buildings in those days is believed to have been the palace of Jerusalem's governor. The area is referenced in historical sources, though its very existence was shrouded in mystery until this excavation, conducted by Israeli Jews, revived it from its dusty grave.[6]

Like a layered cake, the cleared area began to yield up even more long-buried secrets, as Byzantine structures saw sunlight for the first time in twelve hundred years. Archaeology had once more confirmed the obvious: that Palestinian Arabs, far from being the "original" inhabitants of the land, had seized it by force from the ancient empire headquartered in Constantinople. Private dwellings, laden with common household items, bore mute testimony to the personal lives of Jerusalem's Byzantine occupiers. For their part, the Byzantines were hardly the "indigenous peoples" of the region. Digging tenaciously downward,

---

4 Mazar, *Complete Guide*, 162.

5 See W. Harold Mare, *The Archaeology of the Jerusalem Area* (Eugene, OR: Wipf & Stock, 1987), 272.

6 Rivka Gonen, *Contested Holiness: Jewish, Muslim, and Christian Perspectives on the Temple Mount in Jerusalem* (Jersey City, NJ: KTAV Publishing House, 2003), 161.

the excavators discovered, beneath everything else, an elaborately de-signed public space from the days of the notorious King Herod. It was now possible, standing on Herodian flagstones at the foot of the Temple Mount, to envision the grandeur of ancient Jerusalem, ruled by Romans but Jewish to its core. This was the inconvenient truth that would nev-er be acknowledged on the Arab side of the Middle East conflict, the spokesmen of which continued to insist that the great limestone walls had been erected merely to support the mosque above. As declared by Sheikh Ikrema Sabri, "There is not a single stone with any relation at all to the history of the Hebrews."

Temple Mount Al-Aqsa Mosque: Southern Wall Excavations

As the full amount of debris was cleared away, rows of shops came forth from their entombment, having occupied a broad area along the base of the temple platform on several sides. A massive staircase also came into view, leading to the remains of the Herodian period "Hulda Gate" complex — a double and triple gate — that once served as en-trances, via underground ramps, to the upper esplanade, the "Court of the Gentiles." Remarkably, the remains of an enormously sophisticated ritual bath (*mikveh*) complex also came into view, fed by an aqueduct leading all the way from Bethlehem, traversing a distance of over nine kilometers. Like the Pool of Siloam, on the western side of the Temple Mount, these purification pools served the vast numbers of pilgrims as-cending to the holy precincts and in need of ritual cleansing.

Ancient sources relate that on the southwestern corner of the Temple Mount, the priests would signal the beginning of the Sabbath by blowing a horn (*shofar*). Among the debris being cleared during Mazar's excavations, a limestone block emerged, bearing a poignant inscription. Having fallen into the valley from the uppermost portion of the freestanding wall during the Roman destruction of the city, it read, in finely chiseled Hebrew lettering: "To the House of Trumpeting." Some point to this as evidence that Hebrew (not only Aramaic, as often assumed) was in fact a living language in that day. The discovery further bolsters the proposition that the Haram al-Sharif was in fact the Jewish Temple Mount. It dovetails with an unassuming stone inscription in Greek, discovered in 1871, reading:

> No foreigner is to enter within the balustrade and enclosure around the temple area. Whoever is caught will have himself to blame for his death which will follow.[7]

"To the House of Trumpeting"

Senior curator of Hellenistic, Roman, and Byzantine Archaeology, David Menorach, commented, "If we talk about the closest thing to the temple we have on the Temple Mount, this was closest."[8]

Mazar's excavations also shed new light on another feature of the southwest corner of the great retaining wall, known as Robinson's Arch,

---

7 See Peter Connolly, *The Holy Land* (Oxford: Oxford University Press, 1998), 36.

8 https://www.timesofisrael.com/ancient-temple-mount-warning-stone-is-closest-thing-we-have-to-the-temple/.

named after scholar and archaeologist Edward Robinson, who identified its remnants in 1838. Robinson believed that he had identified the eastern edge of a bridge that linked the temple compound with the Upper City of Herod's day. Decades later, however, intrepid explorer Charles Warren (whose sobriquet was "the Mole") discovered the footings of the arch as he dug down to the west of the wall:

> Farther on we mined our way, until, when at a distance of fifty-three feet from the temple wall, we made the grand discovery, which was to throw all controversy on Jerusalem topography into a new groove, to upset many hitherto much cherished theories. It was no less than the pier of that great arch, the springing of which above ground Dr. Robinson had first perceived to be anything more than a projecting stone in the face of the wall, and which he ... supposed to have stretched right across the valley, to the opposite side, joining the upper city to the temple...[9]

Robinson's Arch 1884

Warren concluded: "Now we have ascertained that this arch was not one of a series, reaching across the valley to the Upper City, and so far Dr. Robinson was mistaken."[10] Only during Mazar's new excavations (condemned by UNESCO and much of the international community) was it discovered that the pier was in fact the western support of a single great arch, carrying traffic up from Jerusalem's lower market area

---

9 Sir Charles Warren, *Underground Jerusalem: An Account of Some of the Principal Difficulties Encountered in its Exploration and the Results Obtained* (London: Richard Bentley and Son, 1876), 314.

10 Ibid., 316.

to the Royal Stoa complex, a great colonnaded porch on the Temple Mount's esplanade. Of course Mazar's excavations did not advance the Palestinian narrative that they were the indigenous people of the land, and would not be appreciated, either by the *Waqf* or the international community.

Robinson's Arch Reconstruction

Following the 1993 Oslo Accords, the archaeologically-oriented anti-Israel opprobrium persisted, as members of the Jordanian-controlled *Waqf* (relative moderates in the highly charged Middle East) were replaced with representatives in league with the Palestinian Authority. Thereafter, the Muslim "trust" ceased its cooperation with Israel altogether. Long after the southern wall excavations were completed, in 1995, the *Waqf* decided yet again to manufacture an outrage, erupting in an orchestrated choir of discontent. It was alleged that a new round of excavations had begun in the same sensitive area, potentially threatening the integrity of the Haram.

Another UNESCO representative, a certain Professor Lemair, was dispatched, only to report back that no fresh excavations were being conducted at all. Instead, a reinforcement project was underway, aimed at stabilizing and restoring what had already been excavated. While noting that the excavations had been conducted on privately owned property and without permission of the owner, he declared that this was the most spectacularly important archaeological project taking place in Jerusalem.[11] Of course, this was by no means the first time that private land has been appropriated and co-opted (via "eminent domain" laws) for a larger, public good.

Herodian Quarter

Perhaps the most intriguing site in Jerusalem's Old City came to light through a separate set of excavations. The sites known today as the Herodian Quarter and the Burnt House compromise a subterranean area exposed after the Six-Day War, when the demolished and desecrated Jewish Quarter was being rebuilt from the ground up. Among the underground ruins subsequently uncovered were burned priestly villas, providing evidence of the devastating conflagration that occurred in Jerusalem a month after the Roman destruction of the Temple. The discoveries include spectacular mosaics and artwork, lounges for hosting family and friends, ancient household utensils, and ritual immersion baths. The skeletal arm of a young woman, doubtless killed during the Roman siege, came hauntingly to light.[12] There was even a wall carving, depicting the temple's menorah, said to have been engraved by the

---

11 Mazar, *Complete Guide*, 162.

12 For more on these excavations, conducted by Nahman Avigad, see Jodi Magness, *The Archaeology of the Holy Land: From The Destruction Of Solomon's Temple To The Muslim Conquest* (New York: Cambridge University Press, 2012), 143-147.

high priest himself and quite different from the "traditional" menorah we know from other representations, such as the Arch of Titus in Rome. It is a testament in stone that the Temple Mount is linked across twenty centuries with the city itself, and that it remains the beating heart of the Jewish people.

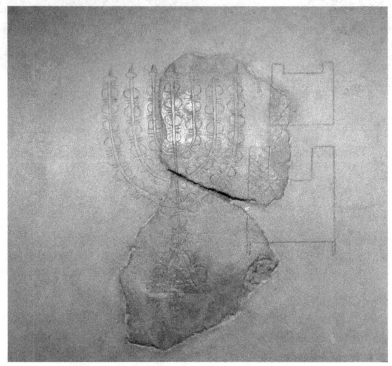

Menorah - Herodian Quarter

Insatiable Condemnation

At the United Nations, however, condemning Israel is habitual, and the multiple seasons of Israel-bashing were insufficient to sate the appetites of the anti-Semitic international community. The extent to which archaeology underscores the historical rights of the Jewish people to live in their ancient homeland had to be met with an endless stream of counterclaims and/or denied altogether. So it was that in 2016 two additional resolutions were adopted by UNESCO, lending its imprimatur to the Palestinian cause, while condemning Israel as "the occupying power" and demanding that it cease and desist from all activities celebrating the Jewishness of Jerusalem. As stated in the "Occupied Palestine Resolution" of October 13, 2016, the Executive Board of UNESCO:

Deeply deplores the failure of Israel, the occupying Power, to cease the persistent excavations and works in East Jerusalem particularly in and around the Old City, and reiterates its request to Israel, the occupying Power, to prohibit all such works in conformity with its obligations under the provisions of the relevant UNESCO conventions, resolutions and decisions...[13]

In response, it might easily be counter-argued that the Israelis are indeed an "occupying power;" one need only dig down a few meters in any part of "Palestine" to recognize that they are occupying their own ancient land. In any case, as far as "oppressed minorities" are concerned, UNESCO could not be bothered with the "facts underground." Nothing must be allowed to interfere with the narrative of victimhood. While acknowledging, in a sentence added later, the "importance of the Old City of Jerusalem and its walls for the three monotheistic religions," the resolution referred to the Temple Mount only as the Haram al-Sharif, effectively writing the holiest site in Judaism out of existence. The Western Wall Plaza was referenced only as the Al-Buraq Plaza, in honor of the mythical steed who whisked the prophet Muhammad on his single-night's journey from Mecca to Jerusalem and back again. Israel was condemned for failing to ensure the exclusive Muslim use of the Haram al-Sharif and was told to put an end to all archaeological excavations in the vicinity of the Old City.

The resolution specifically took aim at the Kedem Center, at the entrance to Silwan, being built to showcase the archaeology of the City of David, along with the Liba House project, designed to highlight previously covered ancient remains by exposing the lowest portions of the Western Wall. It was there that another significant discovery of "unconsidered trifles" was made, in the form of a number of ancient coins. Found beneath the wall's foundations and struck around 17 C.E., they suggest that Herod the Great did not even come close to completing construction on the Temple Mount complex. They confirm the account of Josephus that the holy house and platform on which it stood were not finished until the reign of Herod's great-grandson, Agrippa II, probably around 50 C.E.

Notwithstanding the historical value of such work, the UNESCO resolution:

...urges Israel, the Occupying Power, to renounce the above-men-

---

13 https://unesdoc.unesco.org/ark:/48223/pf0000246215.

tioned projects and to stop the construction works in conformity with its obligations under the relevant UNESCO conventions, resolutions and decisions.

Equally troubling, the resolution emphasizes that the wooden Mughrabi Bridge, ascending from the Western Wall (al Buraq) Plaza to the Temple Mount (Haram al-Sharif) esplanade, represents "an integral and inseparable part of al-Aqsa Mosque," over which the *Waqf* must have "exclusive authority." Should such a step be implemented, it would effectively bar Jews and tourists alike from the Temple Mount.[14] Israeli Prime Minister Netanyahu was quick to respond:

> The theater of the absurd continues at the UN. Today UNESCO adopted its second decision this year denying the Jewish people's connection to the Temple Mount, our holiest site for more than three thousand years. What's next? A UNESCO decision denying the connection between peanut butter and jelly?[15]

Early the following year, at the speech in honor of Holocaust Remembrance Day, António Guterres, Secretary-General of the United Nations, committed the cardinal sin of referring to the destruction of the Jewish temple at the hands of the Romans in the year 70 of the Common Era:

> Imperial Rome not only destroyed the temple in Jerusalem, but also made Jews pariahs in many ways. The attacks and abuse grew worse through the triumph of Christianity and the propagation of the idea that the Jewish community should be punished for the death of Jesus – an absurdity that helped to trigger massacres and other tremendous crimes against Jews around the world for centuries to come.[16]

Such talk was not to be tolerated by the politically correct champions of the "oppressed" denizens of "Palestine." As if he were Galileo, facing the Vatican Inquisition, Guterres was instructed by titular president of the Palestinian Authority, Mahmoud Abbas, to recant of his

---

14 http://www.thetower.org/4088-the-unesco-temple-mount-resolutions-are-even-worse-than-you-thought-heres-why/.

15 https://www.jpost.com/Breaking-News/UNESCO-No-connection-between-Temple-Mount-and-Judaism-470050.

16 https://www.un.org/sg/en/content/sg/speeches/2017-01-27/secretary-generals-memory-victims-holocaust-remarks.

heresy and to issue a formal apology to the people of Palestine. It took an unusual degree of fortitude for Guterres to defy the P.A. He did so in an interview with Israel Radio, insisting (with unmitigated "audacity") that "…it is completely clear that the Temple that the Romans destroyed in Jerusalem was a Jewish temple." He thus succeeded in earning the formal condemnation of the P.A., since he had "ignored UNESCO's decision that considered the al-Aqsa Mosque of pure Islamic heritage" and had "violated all legal, diplomatic and humanitarian customs and overstepped his role as secretary general."[17] The shame of it!

The "Real" Undermining

Meanwhile, as Israel's excavators were being castigated for their hubris in uncovering remains of the ancient structures rimming the Temple Mount, and for "undermining" the Dome of the Rock and the Al-Aqsa Mosque, the *Waqf* initiated an undermining operation of its own. At the southeastern end of the great esplanade, roughly three hundred feet from the mosque, is an approach to a vast underground area, supported by vaulted arches and dating to Herodian times.[18] Legend has it that in the Middle Ages, during the Crusades, the illustrious Knights Templar tied up their horses there, and the space subsequently came to be known as "Solomon's Stables." Beginning in 1996, the *Waqf* launched an extensive construction project, converting Solomon's Stables into a new mosque by adding lights and floor tiles, and renaming it the Marwani Prayer Hall.[19] It was designed to accommodate the additional throngs of worshipers during Ramadan, as well as those who, on rainy days, would otherwise gather on the wide plaza above.

In 1999, the *Waqf* announced its intention to create a small emergency exit for the El-Marwani Mosque. When the Israelis objected, claiming that the *Waqf* planned to exceed its own mandate, a huge vaulted entrance, framed by two great arches, was created instead. In the process, bulldozers were brought in, excavating a trench measuring forty meters in length, with a depth of twelve meters. More than nine thousand tons of dirt and gravel, mixed with ancient artifacts, were removed, with no supervision or inspection by archaeologists. All of it was discarded as refuse into the Kidron Valley below. This archaeologically

---

17 https://www.worldjewishcongress.org/en/news/palestinians-want-new-un-chief-to-apologize-over-jewish-temple-remarks-1-1-2017.

18 See Hershel Shanks, *Jerusalem: an Archaeological Biography* (New York: Random House, 1995) 141-45.

19 See Gonen, op. cit., 167-71.

rich material contained remains of the First and Second Temple periods, but was now rendered all but useless, having been haphazardly mixed together and hauled off in dump trucks. This operation was said to have brought about the most devastating damage ever inflicted on antiquities in Jerusalem.[20]

Moreover, this new construction caused a noticeable bulge in the southern wall itself, endangering its structural integrity.[21] It is deeply ironic that the *Waqf* was guilty of the very thing of which it had accused the Israeli archaeologists. In a compromise negotiation between Israel, the *Waqf* and the Palestinian Authority, the Kingdom of Jordan was assigned to repair the damage. The result, however, was an unsightly patch of bright, light-colored stone, of smooth surface and not in-keeping with the rest of the wall. It has been called a "terrible job" of historical restoration.[22]

The Israeli archaeological community was furious at their own government, which did not interfere at the outset, given the sensitive nature of relations with the Arabs, and the political tensions likely to be exacerbated if action were taken to stop these activities. Historian Eyal Meiron opined:

> That earth was saturated with the history of Jerusalem. A toothbrush would be too large for brushing that soil, and they did it with bulldozers.[23]

The international media were characteristically silent about what could only be described as an archaeological catastrophe. Nonetheless, an archaeology student at Israel's Bar-Ilan University (Zachi Zweig) corralled fifteen volunteers to help gather samples from the mound of refuse, which were then presented to an academic conference. However, the Israel Antiquities Authority (the quintessential government agency that it is) expressed complete disinterest in sifting through the piles of rubble discarded by the *Waqf*. The IAA's chief archaeologist for Jerusalem, Jon Seligman, declared:

---

20 See Aaron Klein, op. cit., 80.

21 See Michael Dermer, *Jerusalem Unbound: Geography, History, and the Future of the Holy City* (New York: Columbia University Press, 2014), 168.

22 See Hershel Shanks, "Temple Mount Repairs Leave Eyesores," *BAR*, 36/5 (2010).

23 Joshua Hammer, "What is Beneath the Temple Mount," *Smithsonian Magazine*, Apr, 2011: https://www.smithsonianmag.com/history/what-is-beneath-the-temple-mount-920764/.

This is nothing but a show disguised as research. It was a criminal deed to take these items without approval or permission.[24]

Though questioned by Israeli police, Zweig's exploits created a media buzz and garnered the attention of noted archaeologist and Bar-Ilan professor, Gabriel Barkai.

In 2004, he and fellow archaeologist Zachi Dvira applied for and received a license from the Antiquities Authority to sort through the mounds of "garbage" dumped into the valley, in the hope of salvaging at least something. Trucks were brought in to remove this "waste" and transport it to a makeshift site at the Emek Tzurim National Park on Jerusaem's Mount Scopus. A large tent-like structure, covered with sheets of plastic, was set up, with a posted sign reading "Temple Mount Salvage Operation." Volunteers of all stripes, from students to international tour groups, were welcomed and encouraged to lend a hand in the meticulous work of sifting through the debris.

The Sifting Project

The tools employed were low-tech, consisting of square wooden frames to which wire screening had been attached. Bucket after bucket of dirt was then emptied into the frames and rinsed with a water hose, the small granules falling through, while the larger pieces, ranging from potshards to ancient coins, were trapped by the screen. While the project's finds have been wrenched out of their archaeological context

---

24 Ibid.

(certainly not *in-situ*), they constitute the *first-ever* archaeological data originating from below the Temple Mount's surface. Among the sponsors of the ongoing project is the same allegedly right-wing "settlers" organization, El'Ad, who have also excavated the City of David. It is another instance of privately funded archaeology doing what government cannot, or, due to political constraints, will not do.

Barkai, along with his full-time cohort of ten, doggedly persisted, hoping to identify, as if by "dumb luck," anything of value among the mass of waste and dross. While most of the buckets yielded nothing more valuable than coagulated lumps of sludge, one of the volunteer sifters would occasionally encounter something of genuine significance. The finds from the First Temple Period range from the tenth century B.C.E. until the destruction of the Second Temple in 70 C.E. For example, three ancient Egyptian or Egyptian-designed scarabs emerged from the muck. There was an abundance of pottery, jugs, juglets, fragmented terracotta figurines, stone sling-shots, shekel stone weights, arrowheads, and other artifacts. A 2,600 year-old seal impression was also discovered, containing three lines of ancient Hebrew, predating the First Temple's destruction in 586 B.C.E. Barkai declared that the find was the first of its kind from the time of King David.

Finds from the Sifting Project

There were also broken pieces of terra-cotta figurines dating from somewhere between the eighth and sixth centuries before the Common Era, possibly bearing witness to the reign of King Josiah, who, we are told, declared nothing short of a war on idolatrous practices within his realm. Several years into the project, a remarkable seal impression was noticed on what appeared as only a black lump of clay. Barkai noted: "It bears the name Gedalyahu Ben Immer Ha-Cohen, suggesting that the owner may have been a brother of Pashur Ben Immer, described in the Bible as a priest and temple official."[25] Gedalyahu was, according to the book of Jeremiah, the chief administrator of Solomon's Temple. An excited Barkai called it the first Hebrew inscription from this time period ever to be discovered on the Temple Mount.

From the Second Temple period, part of a Hasmonean-era oil lamp dating from the second century B.C.E. was uncovered. Numerous Opus Sectile floor tiles from the Temple Mount courts were also discovered among the rubble. These consisted of stones of assorted shapes and colors, fit together in order to display geometric patterns.[26] There were even remnants of bone and ivory combs, most likely used during the purification rituals required before entering a *mikveh* and ascending to the sacred esplanade. Another find of macabre significance consisted of an arrowhead on a shaft, most likely used by Roman legionnaires during the siege of Jerusalem, in 70 of the Common Era.[27]

Additionally, the Sifting Project has uncovered more than five thousand coins, ranging from tiny silver Persian Period coins (from the fourth century B.C.E.) until modern times. The first coin recovered was minted during the Great Jewish Revolt, between 66 and 67 C.E. It bore the phrase "For the Freedom of Zion," and was particularly meaningful, since the Temple Mount was one of the focal points of the fighting. Also found was an extremely rare silver coin, minted during the Revolt. Its face features a branch of three pomegranates and an inscription: "Holy Jerusalem." The reverse of the coin features a cup with the writing: "Half Shekel." Such coins were used to pay the Temple tax during the revolt. Of course, since all the material was jumbled together, without archaeological context, artifacts from any number of time periods might be encountered, from a Crusader coin bearing the image of the Church of the Holy Sepulcher to the badge of a serviceman in the Australian Medical Corps, serving under British general Edmund Allenby during the First World War.

In today's Middle East, however, archaeology has become political.

---

25 Klein, op. cit. 80.

26 "Jerusalem Biblical Temple Floor Designs 'Restored,'" *BBC News*, Sep. 6, 2016: https://www.bbc.com/news/world-middle-east-37288925.

27 Klein, op. cit. 80.

The chief archaeologist of the *Waqf*, Yusuf Natsheh, has declared that all such finds are worthless, having not been found *in situ*. He further declared that the material had been thoroughly examined and that nothing of significance was found. "Every stone is a Muslim development," he insisted.[28] Of Barkai he commented, "This is all to serve his politics and his agenda," namely, that the Jewish connection to the Haram al-Sharif/ Temple Mount is stronger than that of the Arabs. While the word "stronger" may be subject to debate in the minds of some, "older" should be understood as a truism. Some things, it must be stressed, are simply incontrovertible.

"Spoiled Politics"

In October 2007, the *Waqf* commissioned the laying of a new electrical cable in the southeast quadrant of the great esplanade. This involved digging a trench on the Temple Mount some three hundred yards in length. When examined by a team from the Israel Antiquities Authority, a "sealed layer" was discovered, bearing ceramic table vessels such as bowl rims and bases, the rim of a storage jar, the remains of a small oil jug, and a jug handle. Animal bones, possibly the ancient remains of temple sacrifice, were also found, all of which could safely be dated to the seventh and eighth centuries B.C.E. This would place the artifacts within the reign of the biblical King Hezekiah, during the First Temple period. The director of the W. F. Albright Institute of Archaeological Research, Seymour Gittin, observed, "They were pottery of the kind we normally associate with Israelite culture as distinct from the Moabite and other cultures close to Jerusalem at that time."[29]

Since not a trace has ever been found of Solomon's Temple *per se*, the presence of such items in an archaeological layer would at the very least provide circumstantial evidence of its existence. Senior archaeologist Jon Seligman stated, "These finds are important because it is the first time we have ever found a sealed archaeological level clearly dated to the First Temple period within the complex of the Temple Mount." He observed, "We have many finds from the First Temple era from all over Jerusalem, but never from this site." Not surprisingly, Yusuf Natsheh countered the discovery:

---

28 Hammer, op. cit.

29 Matthew Kalman, "Temple Mount Discovery Leads to Dispute in Jerusalem," *SFGate*, Nov. 18, 2007: https://www.sfgate.com/news/article/Temple-Mount-discovery-leads-to-dispute-in-3301191.php.

I was present throughout this work and neither I, nor any Waqf official, recall seeing these items in the trench. I only heard about them in the press, weeks after the work was finished. If they were found, then why were they taken outside the compound?[30]

Might it have been because if the artifacts were not removed, they would never have seen the light of day? Natsheh continued: "All of this archaeology and science in Jerusalem is manipulated for different political attitudes. It is not archaeology, it is not history, it is just spoiled politics." To that Seligman responded, "Categorically, one hundred percent of these findings come from the Temple Mount, and we stake our reputation on that."[31] On at least one account, however, Natsheh is quite correct. When it comes to Palestinian "temple denial," there is plenty of "spoiled politics" to go around.

---

30 Ibid.
31 Ibid.

CHAPTER FOUR

BLOWING UP HISTORY –
PLAYING WITH FIRE AT THE
TEMPLE MOUNT

WHEN MARK TWAIN VISITED the Holy Land in 1867, he found a backwater province of the Ottoman empire, neglected over the course of multiple centuries of inept maladministration. He wrote of his historic visit as follows:

A fast walker could go outside the walls of Jerusalem and walk entirely around the city in an hour. I do not know how else to make one understand how small it is. The appearance of the city is peculiar. It is as knobby with countless little domes as a prison door is with bolt-heads. Every house has from one to half a dozen of these white plastered domes of stone, broad and low, sitting in the centre of, or in a cluster upon, the flat roof. Wherefore, when one looks down from an eminence, upon the compact mass of houses (so closely crowded together, in fact, that there is no appearance of streets at all, and so the city looks solid,) he sees the knobbiest town in the world, except Constantinople. It looks as if it might be roofed, from centre to circumference, with inverted saucers. The monotony of the view is interrupted only by the great Mosque of Omar, the Tower of Hippicus, and one or two other buildings that rise into commanding prominence... It seems to me that all the races and colors and tongues of the earth must be represented among the fourteen thousand souls that dwell in Jerusalem. Rags, wretchedness, poverty and dirt, those signs and symbols that indicate the presence of Moslem rule more surely than the crescent-flag itself, abound... Jerusalem is mournful, and dreary, and lifeless. I would not desire to live here.[1]

He added:

> Renowned Jerusalem itself, the stateliest name in history, has lost all its
> ancient grandeur, and is become a pauper village; the riches of Solo-
> mon are no longer there to compel the admiration of visiting Oriental
> queens; the wonderful temple which was the pride and the glory of
> Israel, is gone… It is a hopeless, dreary, heart-broken land.[2]

Recovering Lost Glory

To appreciate the current, even explosive passions linked to Jerusalem and the Temple Mount, we need to contrast what Mark Twain
witnessed with what it once was. A cursory review of history tells us
that twenty long centuries ago, Jerusalem with its resplendent temple
was an unquestioned and almost unrivaled wonder of the world. Even
while under the symbolic domination of the Roman eagle, displayed
prominently atop the idolatrous standards brought into the city in the
days of the prefect Pontius Pilate, the temple served a totemic function,
underscoring the Jewishness of this city and this land. The temple of
those days replaced an earlier shrine, dedicated in the year 515 B.C.E.
at the hands of Jews returning from seventy years of exile in Babylon
– the "Babylonian Captivity." That "Second Temple," as it was known,
replaced the original Solomonic structure. The biblical account supplies
great detail regarding the construction, the dimensions and the interior
furnishings of the first great temple, which was ultimately destroyed by
King Nebuchadnezzar of Babylon in the year 586 B.C.E. The glory of the
sacred shrine is described in the Bible:

> And it came to pass, when Solomon had finished the building of the
> house of the LORD, and the king's house … the LORD said to him:
> "I have heard your prayer and your supplication, that you have made
> before Me: I have hallowed this house, that you have built, to put My
> name there forever; and My eyes and My heart shall be there perpetu-
> ally." (1 Kings 9:1, 3)

The natural place to look for archaeological evidence of the Holy
House of King Solomon is obviously the Temple Mount/ Haram al-Shar-

1 Mark Twain, *The Innocents Abroad: Or, The New Pilgrims' Progress*, Vol. I (New
York: Harper & Brothers, 1911), 298.
2 Ibid., 607.

if, but no one knows exactly where on the vast plateau the edifice actually stood. Moreover, the religious restrictions imposed by the *Waqf* have made digging in the most promising areas of the great esplanade effectively off limits. In the nineteenth century, however, access was considerably less complicated. In the same year that Mark Twain embarked on his Holy Land pilgrimage, twenty-seven year old British mining expert, Lieutenant Charles Warren, arrived with his own objective. He had been dispatched by the Palestinian Exploration Fund to explore the fortifications of the City of David, the authenticity of the Church of the Holy Sepulcher, and the site of the Solomon's Temple. "The Mole," as he came to be known, recorded the suspicions aroused by his activities:

> An unaccountable rumor arose, that I had arrived on a very sinister mission. That I was to place small packets of gunpowder around the sanctuary walls at a great depth. And then, when they had grown and developed over the course of years into barrels of gunpowder, I was again to come and fire them in order to destroy the grand old walls.[3]

Pressure mounted on the Ottoman authorities, and Warren was subsequently barred from digging anywhere in the vicinity of the Muslim shrine. Not to be deterred, he found another way:

> Any proprietor of land could dig in his own ground, and if I made my bargain with the proprietor, I was really only digging for him. According, I picked out a spot along the south wall, concealed behind some prickly pears, where we worked down along the wall in security... I dug vertical shafts all the way to bedrock, then tunneled perpendicularly to the holy hill. Undetected for months, I was able to map the whole topography of ancient Jerusalem.[4]

To this day, Warren's work remains the best picture of the Temple Mount in existence, though it is but a partial snapshot at best. An inscrutable question remains. Is there anything left of Solomon's Temple to be found, even if prohibitions on digging were not an obstacle? Some scholars, known as "minimalists," have gone as far as to assert that there never was such a temple, that it was nothing more than a contrived fantasy in the minds of later generations of Jews, who wistfully imagined what must have been. Such assertions are more than welcome among

---

3   Linda Osband, ed., *Famous Travellers to the Holy Land: Their Personal Impressions and Reflections* (Eugene, OR: Parkwest Publications, 1989), 146.
4   Ibid.

Palestinian propagandists, intent as they are on nullifying all Jewish connections to this land. There is, to be sure, always something to find amid the archaeological layers, even after destruction and burning, and if the temple had been a reality rather than a Jewishly-manufactured illusion, definite remnants, even foundation stones ought to be there, nestled among the burnt layers.

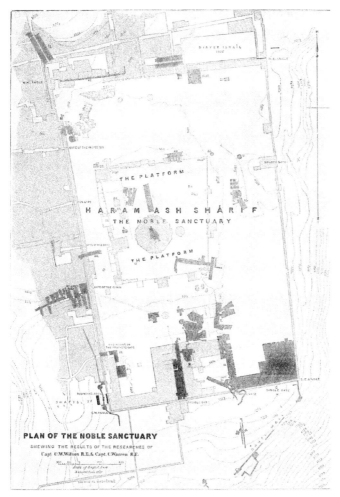

Warren's plan of the Temple Mount

In the 1970s, Professor Benjamin Mazar of the Hebrew University began a collaboration with Dutch-born archaeologist Leen Ritmeyer, to find any evidence of the first great temple of Solomon. Referring to the works of the ancient Greek historian, Strabo, the team spent five years exploring the site. At the lowest course level, they found masonry of a different variety than that of subsequent periods, identifying it as most likely the original footings of the temple. In an account for the Biblical Archaeology Society, Ritmeyer described sections of the rock (directly underneath the golden dome of the Muslim shrine) cut completely flat, which north-to-south have a width of six cubits, precisely the width attributed by the Mishnah (Middot 2:1) to the wall of the Holy of Holies. Ritmeyer proposed that these flat sections constitute foundation trenches on top of which the walls of the original temple were laid.

Well of Souls, beneath the Dome of the Rock

Amazingly, in the floor of the Dome of the Rock was a carved rectangular depression, 2.5 cubits long and 1.5 cubits wide – exactly the dimensions of the Ark of the Covenant (according to the Book of Exodus):

> And they shall make an ark of acacia-wood: two cubits and a half shall be the length thereof, and a cubit and a half the breadth thereof, and a cubit and a half the height thereof. (Exodus 22:10)

Ritmeyer also established that one of the steps ascending toward the Dome of the Rock is quite possibly the top of a stone course from the pre-Herodian Western Wall of the Temple Mount platform. This identification is not without its problems, since, according to Josephus, there were thirty-one steps up to the Holy of Holies from the lower level of the Temple Mount, and according to the Mishnah, twenty-nine steps in total, each step being half a cubit in height. This amounts to a height of at least twenty-two feet — while the height of the *Sakhra* ("rock" in Arabic) is only twenty-one feet above the lower level of the Temple Mount. Nevertheless, the measurements are close enough that serious attention should be paid to what might be the actual spot where Solomon's temple stood and where the Ark once reposed.[5]

Artifacts: Disputed and Undisputed

Additional evidence exists as well, in the form certain artifacts which make reference to the temple, such as the so-called Temple Ostracon, now residing at the Israel Museum of Jerusalem.[6] While only a pottery shard, dating from around 800 B.C.E., it bears ancient Hebrew script featuring the words *Beit Hashem* ("the House of God"). If there had been no temple, why would there be even a single potshard mentioning it? Another ostracon from roughly the same period amounts to an ancient tax receipt for a contribution of three shekels to the "House of the Lord."

Among other artifacts is a small ivory pomegranate vase with a long neck and petals. It dates to around 750 B.C.E., the reign of King Uzziah of Judah. It bears an inscription around its shoulder in paleo-Hebrew script reading: "Sacred donation for the priests of the House of the Lord." It appeared anonymously on the antiquities market in 1979 and was bought in the 1980s by the Israel Museum of Jerusalem for a price tag of $550,000.[7] It was thought to be the most important artifact in the entire collection of biblical antiquities, but allegations later arose that it was in fact a fraud. In 2004, Israeli antiquities collector Oded Golan was indicted for forging antiquities, including the pomegranate, the first time a criminal court had been brought into a case involving archaeological remains. In late 2008, Israeli scientist Prof. Yitzhak Roman concluded that the inscription shows no indication of having been forged. In the end, despite an assortment of scientific testimony against Golan, the consensus of the court was that the prosecution had failed to make its case. While Golan was acquitted in March 2012, it should be noted that the pomegranate was not among the individual counts included in

---

5 Randal Price, *The Stones Cry Out* (Eugene, OR: Harvest House, 1997), 211-17.

6 Ibid., 164.

7 Brian Haughton, *History's Mysteries: People, Places and Oddities Lost in the Sands of Time* (Franklin Lakes, NJ: New Page Books), 57.

the judge's findings.[8] To this day, the matter remains unresolved.

Ivory Pomegranate Inscription

There is, unsurprisingly, yet more evidence to consider. In the summer of the year 2000, another inscribed tablet was said to have been found on the Temple Mount in Jerusalem by the Islamic Trust renovators of the El-Aqsa Mosque. The tablet has subsequently been held by Oded Golan, who was charged with forging it as well. It consists of fifteen lines of Hebrew text, with some elements of Aramaic and old Phoenician. The inscription is, to say the least, significant, as it describes repairs made to Solomon's temple, on the order of King Joash in the 800s B.C.E. The tablet relates how Solomon ordered the priests to "take holy money … to buy quarry stones and timber and copper and labor to carry out the duty with faith." Israel's Geological Institute dated the inscription as approximately 2,800 years old. It was also reported that flecks of gold had been burned into the stone, suggesting that it may have been in the temple at the time it was set aflame by Babylonian invaders in the

8 "Leading Israeli Scientist Declares Pomegranate Inscription Authentic," *BAR* Dec. 16, 2008: https://web.archive.org/web/20100115025132/http://www.bib-arch.org/news/news-ivory-pomegranate.asp.

year 586 B.C.E.[9]

Among the tablet's naysayers was Palestinian archaeologist Hani Nur al-Din of East Jerusalem's Al-Quds University, who suggested that for some, sensationalism and publicity is more important than good scholarship and peer-review. His point is well taken, but we should again note the clear desire among Palestinian Arabs to delegitimize anything that might validate the history of Jewish presence in the land of Israel.

Messages from the Forum

Ironically, one of the most significant of testaments to the historical reality of Jerusalem's most sacred edifice is not in the land of Israel at all, but in the ancient remains of the great city on the Tiber, whose celebrated Forum is the stuff of endless legend. In idolatrous Rome stands a monumental arch, opposite the massive Flavian Amphitheater, known in infamy as the Colosseum. Even as a slumbering ruin, the oval-shaped colossus stands sentinel over the arch in question and the vast archaeological park, which, two millennia ago, was jammed with toga-cloaked citizens of the Roman republic-turned-empire. Begun by the emperor Vespasian on his return from crushing Jewish Zealots in the land of Israel, the Flavian Amphitheater was designed as the world's most grandiose gladiatorial venue. It was the quintessential public space for a grateful Roman citizenry, basking in the glory of having eliminated Jewish resistance to Roman hegemony over its eastern imperial flank.

A structure of this magnitude required many years of murderously brutal labor to complete. As some speculate, the Colosseum may well be the greatest Jewish building ever constructed, having been built in large part by tens of thousands of Jewish slaves transported across the Mediterranean after their defeat in the Great Revolt of 66-70 C.E. They were in all likelihood literally worked to death erecting Rome's most spectacular monument. It was ultimately dedicated by Vespasian's son, Titus, who doused the last embers of Jewish resistance in the land of Israel and returned to Rome in victory. There he was honored with an unrivaled triumphal procession through the Imperial Fora. Telltale impressions in the walls of the amphitheater have allowed contemporary archaeologists to reconstruct what once was a Latin inscription, declaring that the edifice was financed from booty retrieved during the Jerusalem campaign: "Titus, son of Vespasian, built this amphitheater out of the spoils."

---

9   Joseph M. Holden, *The Popular Handbook of Archaeology and the Bible* (Eugene, OR: Harvest House, 2013), 311-12.

Carrying off the Menorah from the Temple depicted on the Arch of Titus

The adjacent Arch of Titus, with its elegantly chiseled reliefs, relates the aftermath of the siege of Jerusalem, as Roman legionnaires carted away the sacred furnishings of the destroyed temple. Among the booty were the Table of the Divine Presence, which stood against a wall in the temple's inner sanctum, and the silver "Trumpets of Truth," which once announced sabbaths and festivals. The most distinctly impressive image on the arch, however, is of the seven branched candelabra, the Menorah, which burned perpetually within the sacred chamber. Josephus writes a vivid description of how these very artifacts, having been transported from holy Jerusalem to the Roman Forum, were paraded through the streets:

But for those that were taken in the temple of Jerusalem, they made the greatest figure of them all. That is, the golden table, of the weight of many talents. The candlestick also, that was made of gold… Its middle shaft was fixed upon a basis, and the small branches were produced out of it to a great length: having the likeness of a trident in their position, and had every one a socket made of brass for a lamp at the tops of them. These lamps were in number seven; and represented the dignity of the number seven among the Jews. And the last of all the spoils was carried the law of the Jews. After these spoils passed by a great many men, carrying the images of victory: whose structure was entirely either of ivory, or of gold. After which Vespasian marched in the first place: and Titus followed him.[10]

To this day, the Arch of Titus bears haunting witness to that great procession and to the presence in Jerusalem of the Jewish temple, not as a dimly remembered phantom structure, but as a tangible reality and the central focus of all Israel. From the moment the great sanctuary came down in flames, there burned in the hearts of Jews everywhere an inextinguishable longing to see it rise again.

Echoes Across Centuries

The Jewish desire to rebuild the hallowed shrine is all the more poignant, in light of the utter grief expressed by those who had experienced the catastrophe firsthand. Known as "Mourners of Zion" (*Avelei Tziyon*), they were recorded as saying:

Blessed is he who was not born
Or he, who having been born, has died.
But as for us who live, woe unto us,
Because we see the afflictions of Zion,
And what has befallen Jerusalem…
And let not the brides adorn themselves with garlands;
And, ye women, pray not that ye may bear…
Or why, again, should mankind have sons?
Or why should the seed of their kind again be named,
Where this mother is desolate,
And her sons are led into captivity?[11]

In seeming answer to their heartbreak, the central prayer of the Jew-

---

10   Josephus, *War,* VII, V, 5, Whiston trans.
11   *Syriac Apocalypse of Baruch* 10:6-7; 9-10-13-16.

ish faith (the *Amidah*), recited three times daily from antiquity, includes these words:

> May it be your will, O my God and God of my fathers, that the Temple be rebuilt speedily in our days, and give us our portion in your Torah, and there we will worship you with reverence as in ancient days and former years.

As far back as the early rabbinic (*Tannaitic*) period, a specified sum of money was regularly put aside for the temple's reconstruction. Only after many years, when it became clear that the Holy House would not imminently be rebuilt, were these monies reallocated for charity and community service. In the Middle Ages, the great Jewish poet of Spain, Yehuda HaLevi, echoed the pathos felt by Jews worldwide over the ruins of the holy sanctuary. In his melancholy "Ode to Zion," he emotively declares:

> The air of your land is the very life of the soul, like myrrh are the grains of your dust; your streams are like the honeycomb. It would be pleasant for me to walk naked and barefoot among your desolate ruins, where once your temples stood, where the ark was hidden, and where your Cherubim dwelled in your innermost shrines.[12]

All the lamentation notwithstanding, the idea of a reestablished sacred shrine has echoed through the centuries into the present. Not surprisingly, arguments and opinions about the exact location of the Holy House on the great esplanade have varied widely from Warren's time to the present. It has in fact been suggested that rather than occupying the spot where the Dome of the Rock currently stands, it may have been situated to the north of the Muslim shrine, immediately opposite the Eastern, or Golden Gate of the Temple Mount. Though walled up by the Ottoman sultan Suleiman the Magnificent in 1541, it may preserve the location of the ancient Shushan Gate, which was the main approach to the temple two millennia ago, in the Herodian era. Others have argued that the sanctuary may have been to the south of the Dome of the Rock, or even to the south of the Temple Mount itself. Any of those options would mean that a third Jewish temple could theoretically be constructed without touching the Muslim shrine at all.

---

12 Lavinia Cohn-Sherbok, Dan Cohn-Sherbok, *Medieval Jewish Philosophy: An Introduction* (London: Routledge, 2013), 64.

American academic Robert Eisenman passionately opined in a 2009 article in the *Jerusalem Post* that Israel had made a mistake when, after conquering the Old City in 1967 under the illustrious General Moshe Dyan, it failed to exert full authority over the Temple Mount:

> Moshe Dayan was wrong in ordering the Israeli flag taken down …
> surrendering sovereignty and giving the Muslim *Waqf* control over
> the Temple Mount. No self-respecting people after two victorious
> wars would have behaved in this way. But he had no guideposts to rely
> upon, only egocentrism and his own pragmatism – plus he loved "the
> grande geste."[13]

Eisenman went on to argue that a new, Third Temple, should indeed rise again in Jerusalem as a living "symbol" for Jews worldwide:

> But now two — three generations after the Holocaust … we have noth-
> ing positive to appeal to our young generations of Jews in Israel and
> abroad. It is poetry and the Spirit that provide this. They are the pos-
> itives, not humiliating renunciations. The reconstruction of a Temple
> – any Temple – should have begun forty years ago and we would be
> well on our way towards achieving these things. This does not mean
> we should emulate the old design. Its content, shape, and operation
> should be open to investigation, even architectural competitions, and
> creativity; but the symbol would be there.[14]

Plots and Plans: "Demolition Derby"

Some in Israel have never been content to harbor wistful dreams of restoring ancient glory, instead embracing a much more radical, even violent agenda. From the moment Israel conquered Jerusalem, Judea and Samaria in 1967, there was much going on, both savory and unsavory, behind the scenes. It took decades before it came to light that the chief rabbi of the Israel Defense Force had in fact advocated blowing up the Dome of the Rock. The shocking revelation came from war hero Uzi Narkiss, who had, during the Six-Day War, led the charge on the Old City of Jerusalem. Narkiss recounted that Rabbi Shlomo Goren had approached him in the hours following the capture of the Temple Mount, urging him to place one hundred kilograms of high explosives inside the

---

13 Robert Eisenman, "Remember, the Temple was Built by Herod," *Jerusalem Post*, Oct. 27, 2009: https://www.jpost.com/Blogs/The-Eisenman-Line/Remember-the-Temple-was-built-by-Herod-367256.

14 Ibid.

golden shrine. Narkiss reported the conversation as he remembered it:

> I said to him, "Rabbi, enough." He said, "Uzi, you will go down in history if you do this." I answered, "My name will already be written in the history books of Jerusalem." But Goren persisted. "You don't grasp what tremendous significance this would have. This is an opportunity that can be taken advantage of now, at this moment. Tomorrow it will be too late." I said "Rabbi, if you don't stop, I'll take you to jail." Thus the discussion, which only lasted a few minutes, came to an end. Rabbi Goren turned and walked away in silence.

Dome of the Rock interior

In fairness, it should be stressed that the details of Narkiss' conversation with the rabbi are disputed. Nonetheless, the story reverberates with an air of veracity, given the transcript of a speech Goren gave to a military convention, also in 1967, in which he advocated bringing down the al-Aqsa Mosque (also known as the Mosque of Omar). Said Goren:

> I told this to the defense minister [Moshe Dayan] and he said, "I understand what you are saying, but do you really think we should have blown up the mosque?" and I said, "Certainly we should have blown it up." It is a tragedy for generations that we did not do so... I myself would have gone up there and wiped it off the ground completely so that there was no trace that there was ever a Mosque of Omar there.[15]

The sentiments expressed by Rabbi Goren were echoed multiple times in the years that followed, coupled with serious plans to put them

---

15 See Hilary Appelman, *Haaretz*, and the AP: "Jerusalem," Dec. 31, 1997.

into effect. A little more than three decades ago, an Israeli with radical designs named Yehuda Etzion became involved in a dark conspiracy that may indeed have resulted in blowing up the Dome of the Rock. Ultimately convicted and imprisoned by the state of Israel for his own brand of Jewish terrorism, Etzion had become inflamed over Prime Minister Menachem Begin's peace deal, whereby Israel gave back the entire Sinai peninsula to Egypt's Anwar Sadat. As an early leader in the movement to settle Judea and Samaria – amounting to a slow annexation of the territories – he joined forces with the so-called "Jewish Underground," to enact their grizzly deed. During his service in the Israel Defense Force, he had learned to handle high explosives, which he decided to use in the service of his convoluted vision.

In the aftermath of a horrific 1980 ambush and murder of six Jewish students in the predominantly Palestinian city of Hebron, the Jewish Underground set out on a course of revenge. Etzion explained:

> In Jewish tradition the period of one month has a special meaning. It's the period of mourning. So we decided on a timetable of one month.

That was when they struck, planting explosives in the cars of three Palestinian mayors. Bassam Shaka, the mayor of Nablus, lost both of his legs from the blast. It hardly mattered that the Shin Bet, Israel's internal security service, found no involvement on the part of the victims in the killing of the students in Hebron. As Karim Gilon, head of the Shin Bet, had said, "We have no information that they were involved in any terror act." However, the case remained unsolved, prompting an intensive investigation that spanned two years, involving ninety policemen and spearheaded by Israel's Serious Crimes Division.

Three years later, the Jewish Underground struck again, in retaliation for the killing of a yeshivah student in Hebron. Two men went into Hebron's Islamic College and opened fire, lobbing a hand grenade as well. Three Muslim students died in the attack, and thirty-three others were wounded.[16]

Armageddon

Meanwhile, Etzion and his co-conspirators were planning "Armageddon." The six hundred pounds of high explosives they had diverted from I.D.F. stockpiles would be sufficient to accomplish the task. As Etzion explained, "Redemption without the Temple is like trying to revive

---

16 Gorenberg, op. cit., 128–137.

someone without a heart." Destroying the Dome, he believed, would accomplish the dual task of undoing the "treacherous" peace with Egypt and clear the sacred plateau so that the Temple may be rebuilt. It would also bring about unimaginable strife in the region. Shekh Muhammad Hussein declared, "Damaging the holy shrine would lead to repercussions, the scale of which I can't even imagine." One cannot overestimate the incendiary nature of the thirty-five acres that comprise the Temple Mount, and the potential of this sacred turf to bring on some kind of apocalyptic conflict throughout the Middle East, even culminating in an "Armageddon" scenario. Izar Beh of Israel's Keshev Center (monitoring extreme nationalist and religious groups) stated: "To harm the mosque means a global war between the Arab world and the Islamic world against Israel, and there is no doubt that it could be a war that may bring destruction to the state of Israel."

Notwithstanding the determination of the Jewish Underground, agents of Shin Bet apprehended fifteen conspirators in April 1984, for planting bombs beneath six busses in Arab East Jerusalem. They were meant to be detonated following Friday prayers, coinciding with the celebration of the *Isra* and *Mira'j* (the Prophet Muhammad's night journey from Mecca to Jerusalem and back), as worshippers were returning from the mosque. In the sting that followed, the settlement of Kiryat Arba (near Hebron) was raided, uncovering significant amounts of high explosives, along with weaponry. There were at least twenty-five arrests, including Rabbi Moshe Levinger and settler leader Eliezer Waldman.[17] As the "big picture" emerged, the most incredible intrigue of all was revealed – the Jewish Underground's targeting of the Dome of the Rock itself.

The guilty verdict that followed on the heels of the trial was a surprise to no one. As if the Middle East were lacking sufficient reason for a cataclysmic war, did Israel really need this as well? For having planned an attack against the Islamic college, three of the conspirators, Menachem Livni, Shaul Nir and Uzi Sharbav, were given life sentences. After serving seven years, Israeli President Chaim Herzog commuted their sentences, and they returned home to heroes' welcomes. While the Jewish Underground hailed them, the larger settlement movement – Gush Emunim ("the Block of the Faithful") – turned on them, and their tactics of killing innocents.

In tandem with such attempts to foment Jewish terrorism, there

---

17 See Robert I. Friedman, "In the Realm of Perfect Faith," *Village Voice* 12 (Nov. 1985).

have been other, equally dark conspiracies. In October 1982, Yoel Lerner, a member of the militant "Kach" movement (founded by Meir Kahane, who was banned from Israel's Knesset as an extremist), was arrested by Israeli police. He was accused of hatching yet another conspiracy to blow up the Dome of the Rock. In the grip of religious fervor he declared: "Since the Romans destroyed the Second Temple, it's as if Judaism has had its heart extracted and is living on borrowed time." Though convicted and sentenced to two-and-a-half years in prison, this would not be the only time Israel would find itself battling its own people. In March 1983, Israeli security forces captured ten Jewish extremists carrying rifles, hoes and crowbars near an ancient passageway to the Temple Mount. Their aim was to perpetrate an armed raid on the Haram al-Sharif and seize Muslim and Jewish holy places.

In January 1984, the most ambitious plan of all to assail the Temple Mount was launched by the so-called "Lifta Band," whose ultimate desire was not only to blow up the Muslim shrines, but to usher in the Messiah. While Israeli security thwarted the plan at the last minute, the nightmare scenario attracted international attention. Israel's Supreme Muslim Council issued a warning: "If the attempted explosions had succeeded all Arab countries would have immediately launched a holy war against Israel."[18] But where, among the community of nations, was there any acknowledgement of Israel's determined efforts to foil the violent aims of radical elements of its own population? Have there been similar efforts on the part of the Palestinian Authority to restrain its homespun radicals? The question is obviously rhetorical. Indeed, the Palestinian Authority names streets in their honor and financially subsidizes the families of suicide bombers.

While it is clear through all of this that the Jewish state has its extremists, it is equally clear that Israel arrests its home-grown terrorists. Israel has certainly done its best to restrain and incarcerate violent and unseemly societal elements. What is less certain is whether the success of Israel's internal security service can be guaranteed forever. It only takes is a single incident, a terrorist strike powerful enough to bring down the Dome of the Rock or the al-Aqsa Mosque, and all bets are off as to the nature and character of the consequences.

The Holy House and the Heifer

In the meantime, an organization called the Temple Institute has

---

18 Alan Balfour, *Solomon's Temple: Myth, Conflict, and Faith* (Chichester: John Wiley & Sons, 2012), 272.

undertaken to re-create the sacred vessels and furniture of the holy house, the ultimate objective being nothing less than the rebuilding of Israel's most sacred shrine atop the Temple Mount. More than ninety items of ritual use in the ancient sanctuary have been identified, and many of these have already been remade. The priestly vestments worn by those conducting animal sacrifice and performing sundry worship rituals have also been fashioned. These include the golden crown and garments of the high priest (*kohen gadol*), down to the "tassels" (*tzitzit*), as well as his sacred breastplate and *ephod*, as described in the book of Exodus.

The work of the Temple Institute has garnered not only fascination, but financial contributions from inside Israel as well as beyond. The burden to rebuild the temple spread, surprisingly enough, to non-Jews in the United States. A Christian preacher and cattle breeder in Canton, Mississippi, by the name of Clyde Lott, was perusing the Scriptures one evening when he noticed that in order for proper purification rites to be performed within the precincts of a third temple, it would be essential to have red heifers on hand and ready for sacrifice. The biblical text is clear:

> This is the statute of the law which the LORD has commanded, saying: Speak to the children of Israel, that they bring you a red heifer, faultless, in which there is no blemish, and upon which a yoke never came. (Numbers 19:2)

Lott immediately made contact with the Mississippi State Trade Office, only to discover that no red bovines existed in modern Israel. Thereafter, in conjunction with the Temple Institute and an assortment of American Christian benefactors, he set out to develop a breed of red heifer that he believes will contribute to the spiritual restoration of the Holy Land. Many years later, on August 28, 2018, Rabbi Chaim Richman, director of the Temple Institute, announced the birth of the first completely red heifer in the land of Israel in the last two millennia. After thorough examination by rabbinical authorities, the cow has been declared a kosher candidate for the biblical designation. Such a biological anomaly well suits the apocalyptic mentality of many evangelical Christians, who believe that since the Jewish people have returned to their homeland, the temple must next be rebuilt, prior to seven years of tribulation, the battle of Armageddon, and the return of Jesus on the day of judgment. All of this curiously hinges on a red heifer – one of the few things on which Christians and temple-obsessed Israeli Jews concur.[19]

While the majority of the Israeli populace call themselves "secular" and are thoroughly disinterested in upsetting the status-quo on the Temple Mount, the vision of a rebuilt sanctuary even has the support of some government ministers. Nonetheless, given the fact that Israel did not assert full civil authority over the great esplanade but instead entrusted the *Waqf* with such oversight, any realistic attempt at reconstructing a Third Temple would certainly result in massive civil unrest in the Arab sector, such as has never been seen before. Even without touching a single stone of the Islamic structures, a new Jewish temple on the Haram al-Sharif would doubtless be accompanied by unprecedented international condemnation, and quite possibly a regional war. A large majority of Israelis feel that the Jewish state has enough reasons to be concerned about the outbreak of war without adding fuel to the already smoldering fires via the construction of a Third Temple.

CC BY-SA 4.0 Edmund N Gall

Replicated Menorah next to the Temple Institute

Even among orthodox Jews there is no agreement about whether the temple may be rebuilt at all until the Messiah comes. Some point to the great medieval expositor of Jewish law, Moses Maimonides, who, they believe, formally condoned the reconstruction of the Holy House. Oth-

---

19 See Gorenberg, op. cit., 7-29.

er rabbinic authorities argue that Maimonides never sanctioned the rebuilding of the temple in absence of the Messiah. There is also the issue of whether observant Jews are allowed in any case to ascend the Temple Mount, since the precise location of the sacred shrine and the Holy of Holies cannot be ascertained with certainty. It is not impossible that a Jew who dares even to walk on the great plateau might inadvertently be treading above the remains of the most holy place. This, however, is not a concern of the Temple Institute and its faithful adherents. They contend that the location of the ancient temple's most sacred inner chamber is in fact known, and that (as Leen Ritmeyer demonstrated) it once occupied the exact spot where the Dome of the Rock stands today. There is therefore no risk that it might inadvertently be defiled.

Ready to Rebuild

The work of the Temple Institute has dovetailed with that of another group, established in 1967 and known as the Temple Mount and Land of Israel Faithful Movement. Its founder, Gershon Salomon, formerly an officer in the Israel Defense Force, was motivated by sheer religious zealotry, and went as far as to dispatch one of his sons to identify the location of Israel's ten "lost tribes" and guide them back to the Holy Land. Also known simply as the "Temple Mount Faithful," they became the first substantial movement in Israel to advocate openly for the reassertion of Jewish authority and civil control over the holy esplanade. For better or for worse, archaeology had without question become geopolitical. Salomon declared: "Whoever controls the Temple Mount has rights over the land of Israel."[20]

As if to underscore the point, in October 1990, the Temple Mount Faithful announced their intention to lay the cornerstone for the proposed Third Temple atop the great plateau. They had gone to the trouble of fashioning the stone itself, and they proceeded to move it into the Western Wall Square. Israel's High Court had already barred the group from entering the Temple Mount itself, since they had on five previous occasions attempted to demonstrate at the site of the al-Aqsa Mosque. Nevertheless, their announcement alone was enough to trigger a violent response from the Palestinian Arabs. Massive rioting begin at precisely 10:30 a.m. on October 8, which came to be known as "Black Monday."

---

20 Eliezer Don-Yehiya, "The Book and the Sword: the Nationalist Yeshivot and Political Radicalism in Israel" in *Accounting for Fundamentalisms: The Dynamic Character of Movements*, Martin E. Marty, and R. Scott Appleby, eds. (Chicago: University of Chicago Press, 2004), 280.

An Israeli inquiry on the violence later reported:

> The members of the *Waqf* knew that the High Court had refused the Temple Mount Faithful petition to lay the cornerstone of the Third Temple, and did not respond to requests by Israel Police officers on the morning of the incident to calm the crowd. This, even after the police informed the Wakf that they would also prevent the Temple Mount Faithful, and anyone else, from visiting the area, though such visits are allowed by law.[21]

The report details what transpired next:

> The incident itself began when, suddenly, violent and threatening calls were sounded over the loudspeakers *"Allahu Akbar"* [God is Great], *"Ahad"* [Holy War], *"Itbah Al-Yahud"* [Slaughter the Jews]. Immediately afterwards, enormous amounts of rocks, construction materials and metal objects were thrown at Israeli policemen who were present at the site. Many in the incited, rioting mob threw stones and metal objects from a very short range, and some even wielded knives. The actions of the rioters, and certainly the inciters, constituted a threat to the lives of the police, the thousands of worshippers at the Western Wall and to themselves. This was a serious criminal offense committed by masses who were incited by preachers over loudspeakers, and this is what led to the tragic chain of events... It is the opinion of the commission that ... a considerable percentage of the people gathered on the Temple Mount and their leaders were involved in the disturbing of public order, causing harm to police and worshippers and endangering their lives... Nineteen policemen were injured as well as nine Western Wall worshippers. According to Police statistics, twenty people were killed and fifty-two injured on the Temple Mount.[22]

A U.N. Resolution followed, proposed, surprisingly enough, by the United States, in which the Security Council declared that it:

> 1. Expresses alarm at the violence which took place on 8 October at the Al Haram Al Shareef and other Holy Places of Jerusalem resulting in over twenty Palestinian deaths and to the injury of more than one hundred and fifty people, including Palestinian civilians and innocent

---

21 *165 Summary of a Report of the Commission of Inquiry into the Events on Temple Mount on 8 October 1990- 26 October 1990*: https://mfa.gov.il/mfa/foreignpolicy/mfadocuments/yearbook8/pages/165%20summary%20of%20a%20report%20of%20the%20commission%20of%20inqui.aspx.

22 Ibid.

worshippers;

2. Condemns especially the acts of violence committed by the Israeli security forces resulting in injuries and loss of human life;

3. Calls upon Israel, the occupying Power, to abide scrupulously by its legal obligations and responsibilities under the Fourth Geneva Convention, which is applicable to all the territories occupied by Israel since 1967;

4. Requests, in connection with the decision of the Secretary-General to send a mission to the region, which the Council welcomes, that he submit a report to it before the end of October 1990 containing his findings and conclusions and that he use as appropriate all the resources of the United Nations in the region in carrying out the mission.[23]

Of course the United Nations cannot deign to refer to Israel as a nation (only as the "occupying Power"), since its very existence remains in dispute and is flatly rejected by a sizable portion of the Arab population and its leadership. The Israeli government, while rejecting the U.N. resolution, made sure to ban Gershon Salomon from ascending the Temple Mount. The Temple Mount Faithful have nonetheless persisted in seeking access to the holy esplanade whenever there is a major Jewish festival. The Israeli authorities regularly turn down these petitions, whereupon the group immediately appeals to the High Court of Justice. The appeals are routinely granted, on condition that the police determine that such actions would not inflame the security situation. While the group would be allowed entrance to the Haram al-Sharif, they would not be allowed to pray there. Only in Jerusalem, the holy "City of Peace," are such things imaginable. While it would take a high court ruling to allow access to the Temple Mount, by the same token, if one happens to be Jewish, one had better not pray there, notwithstanding the biblical admonition: "Pray for the peace of Jerusalem." Muslim prayer by contrast is perpetually protected. The police, for their part, have never arrived at a determination favorable to the Temple Mount Faithful, and block them from ascending to the Mughrabi Gate. On a practical level, the group has been relegated to a protest movement.

Meanwhile, the former director of the *Waqf,* Sheik Adnon Husseini, made a telling statement at the Al-Aqsa Mosque, while dedicating a new podium (to replace the one burned in a fire set by an Australian tourist in 1969). He declared:

This historic occasion proves that the extremist Jews will never achieve

---

23 Resolution 672 (Oct. 12, 1990).

their goals of taking over the Temple Mount. It shows that we are much closer to liberating the al-Aqsa Mosque in Jerusalem from Israeli occupation.[24]

For the sheik, "liberating" most certainly meant the complete withdrawal of Israeli authority and control over East Jerusalem, and one day over all of Jerusalem, indeed the whole land called "Palestine," from the Jordan River to the Mediterranean.

Footloose on the Holy Hill

Such assertions were tested on September 28, 2000, when Ariel Sharon, at the time serving as Israel's Minister of Foreign Affairs, visited the Temple Mount in Jerusalem. His visit ignited a violent revolt from the Palestinians, which started the second "Intifada," or "Uprising," also dubbed the al-Aqsa Intifada. As in the first Intifada, which broke out in November 1987, Palestinians, including young men and boys, began throwing stones and rocks at Israeli troops, causing severe injury to many. The Israeli army responded with tear gas and rubber bullets, only to be condemned in the world community, especially at the United Nations. Both Intifadas had rallied the Palestinian people, but the PLO also realized that this method of opposing the Israelis would not be enough to weaken Israel's resolve or cause it to withdraw unilaterally from the territories won in 1967.

Another ardent campaigner in the ongoing struggle for Jewish access to Israel's holiest site is Knesset member and Orthodox rabbi Yehuda Glick. Having served as executive director of the Temple Institute, he subsequently founded several organizations, such as the Temple Mount Heritage Foundation, in order to draw attention to the need for human rights for people of all faiths on the sacred plateau. Having immigrated from the United States to Israel, he now argues that the exclusive control of the Temple Mount by the *Waqf* is a prime example of discrimination on the basis of religion, at a place which ought to be a house of prayer for all nations.

The proclamation of such ideas in Jerusalem can, however, make one a "marked man." After giving a speech at a conference at Jerusalem's Menachem Begin Heritage Center on October 29, 2014, a motorcyclist spotted Glick as he was loading his car, rode toward him and said, with a thick Arab accent, "I'm very sorry, but you're an enemy of al-Aqsa, I

---

24 Klein. op. cit., 81.

have to."[25] He drew a pistol and proceeded to shoot Glick four times in the chest before making a hasty getaway.[26] Glick miraculously survived the incident, while his attacker, a member of the terrorist group Islamic Jihad, was tracked down by Israeli police and killed in a gun battle.[27]

Golden Gate of the Temple Mount

Yehuda Glick, for his part, foresees the peaceful reconstruction of a Jewish altar of sacrifice, alongside the Dome of the Rock, not in place of it. To highlight the seriousness of his vision, two lambs were ritually slaughtered at the site on March 26, 2018.[28] Glick has repeatedly escorted Jewish groups to the holy hill, walking about freely and praying, until being arrested. In protest to a police injunction on his activities, he initiated a hunger strike in October 2013. The ban was lifted after twelve days, allowing Glick and his friends to ascend to the site, while forbidding them from praying. An Israeli court ruling (arising from an allegation that Glick had pushed and shoved a Muslim woman, who was a guard on the Haram al-Sharif) later affirmed that Jewish prayer at the

25 Yair Ettinger, "Gunman Apologized Before Shooting, Glick Tells Rabbi," *Haaretz*, Nov. 17, 2014.

26 Lazer Berman, "Right-Wing Activist Shot, Seriously Hurt Outside Jerusalem's Begin Center," *Times of Israel*, Sep. 29, 2014.

27 Noam (Dabul) Dvir, "Police Kill Suspect in Shooting of Right-Wing Activist Yehuda Glick" *Ynetnews*, Oct. 30, 2014: https://www.ynetnews.com/articles/0,7340,L-4586020,00.html.

28 Udi Sheham, "Activists carry out Passover ritual sacrifice at the foot of the Temple Mount," *Jerusalem Post*, Mar. 26, 2018.

holy esplanade is in fact legal.[29] One might find it odd that the right to pray might require the imprimatur of a court, but such are the skewed realities on the Temple Mount.

Glick has been barred repeatedly from entering the site, so as not to inflame the delicate status quo vis-à-vis the *Waqf*, but his devotion to Jewish religious freedom on the holy hill remains unshaken. On February 19, 2020, Glick was confronted yet again on the Temple Mount by Israeli police, who accused him of attempting to diverge from the regular "visitors' path" and of refusing to follow their orders. He sat down on the pavement in protest, and was promptly arrested and dragged away to a police station in the Old City. On his release he was accused of absconding with the official papers of his arrest, and an investigation into his conduct was immediately launched.

In the final analysis, Glick's hopes and dreams appear to dovetail with those of Professor Eisenman, who poignantly observed:

> It took the Herodian Temple almost ninety years to be completed. Ours and even its early stage – archaeological investigation – hasn't even begun. The Holocaust Generation is almost past or passing. People need a Positive Historical Judaism to go forward… People need positive symbols to rally around. The time is late. In no other manner can we gain the respect of the world and regain our own self-respect; and the world come to understand us – and we come to understand ourselves.[30]

---

29 Renee Ghert-Zand, (November 18, 2013). "One Man's Jewish Temple Mount Crusade," *The Forward*, Oct. 30, 2014: https://forward.com/opinion/187835/one-mans-jewish-temple-mount-crusade/.

30 Eisenman, op. cit.

CHAPTER FIVE

ISLAMIZING HISTORY –
HEBRON AND THE TOMB RAIDERS

F ROM A DISTANCE the thirsty hills of Hebron appear as a collage of jumbled boxes, a kaleidoscope of dreary pastel shapes, consisting of concrete and stone faced homes, shops and businesses, huddled tightly together amid a patchwork of awkwardly inclined streets. Situated some thirty-two kilometers due south of Jerusalem, it sprawls across an assortment of hills and wadis in the bosom of the territory known from hoary antiquity as Judea. At an elevation of nine hundred thirty meters above sea level, it is also the highest city in the land of Israel.

Conveniently nestled at the intersection of ancient trade routes, Hebron was where caravans and camel trains converged, having criss-crossed the Sinai, Transjordan and the Arabian Peninsula. As with so many Middle Eastern towns, the history of its habitation stretches across multiple civilizations and cultures, from ancient times through the medieval Mamluk and later Ottoman periods, producing a polyglot collection of architectural curiosities. Greatly in evidence is the heavy use of local limestone, much of which is deeply weathered and gray, as if to match the general angst of present day life in this sacred city of biblical lore.

The humming heart of Hebron is its marketplace, its *shuk*, where tireless crowds rummage through table-lined sidewalks, buying cloth-ing, phones, children's toys, food items and more. Eager vendors noisily call out prices, expertly luring customers into their dingy shops. While most of the men dress in contemporary western garb, the elderly are fes-tooned in checkered *keffiyehs*. The women wear head scarves and long dresses. There are no movie theaters, no bars, no nightlife, as the town is in a real sense captive to its conservative religious ethos. The root behind the word "Hebron" means "friend," but in the tinderbox of today's Middle East, friends are hard to come by, and enemies are ever at the gate.

The single landmark to which every eye gravitates is a vast rectangular limestone monument, towering above the surrounding midget structures like a lone giant in a field of Lilliputians. To the Jews it is the Tomb of the Patriarchs; to the Muslims it is known as Haram al-Khalil – the "Shrine of the Friend" [of God].[1] It is one of the oldest places of worship on earth never to have fallen out of use.[2] It is also a prime example of the expropriation and Islamization of Jewish sites by Muslims over the course of many centuries.

Tomb of the Patriarchs

Each of its enormous ashlars is an artistic masterpiece, seamlessly joined to the next in mortarless precision. Its staggering dimensions alone testify to the sacredness of this ground, as if to cry out that something of great importance lies here. Indeed it does. It strains credulity to imagine that a structure so magnificent should rise majestically above a

---

1 Hebron was renamed by Muslim armies in the wake of their conquest of Palestine in 632 C.E. See Edward Platt, *City of Abraham: History, Myth and Memory: A Journey through Hebron* (London: Picador, 2012).

2 Ibid.

subterranean feature as inglorious as a cave. More precisely, it is a double cave, biblically known by the name Machpelah, and believed by both
Jews and Muslims to be the burial place of Adam and Eve, along with
the patriarchs and matriarchs, Abraham, Isaac, Jacob, Sarah, Rebecca
and Leah.[3] Notably, the word "Machpelah" in Hebrew means "double."
Josephus, in the first century, lent his personal imprimatur to the site's
location:

> Their tombs are pointed out to this day in the little town, of the finest
> marble and beautifully fashioned. Three quarters of a mile from the
> town can be seen an immense terebinth, said to be as old as creation.[4]

After the Temple Mount, it is the second holiest site in the Jewish
faith, for which reason the tyrannical King Herod the Great ordered
its construction. Its colossal walls are perfectly aligned toward the four
points of the compass; but as massive as they are, they somehow appear
lighter, via a cleverly deceptive optical illusion. Each stone course is set
back exactly 1.5 centimeters from the one below, and each upper course
is progressively wider, creating the impression of perfectly straight and
symmetrical lines, rather than slowly receding from one's line of sight.
An architectural wonder, it is the only intact Herodian structure still in
existence. The eye is drawn upward, from the solid base of the monument to its great, freestanding walls, with alternating recessed segments
punctuated by faux, squared-off columns that lend a balanced elegance
to the whole. Within those walls is an elevated courtyard, from which a
maze of corridors, passages and chambers branch off, having been appended to the original over twenty centuries.

We know from ancient sources that this gargantuan tomb complex
is nothing less than a mini-representation of what Jerusalem's Temple
Mount once looked like, before its own freestanding walls were leveled
by Rome's legions. The great retaining walls of the temple's vast artificial
plateau – including the Western Wall – are akin to the lower courses
of the Tomb of the Patriarchs. Above them would have stood the same
style of faux columns that still grace the Hebron shrine, to which every
modern representation or model of the Temple Mount owes its inspiration.

---

3 See Ron E. Hassner, *War on Sacred Grounds* (Ithaca & London: Cornell Univ. Press,
2009), 73.
4 G. A. Williamson, trans., *Josephus, The Jewish War*, IV, 533 (New York: Penguin,
1981).

The Biblical Backstory

Hebron of course features prominently in the biblical narratives, going back to the father of the Jewish people, Abram:

> And Abram moved his tent, and came and dwelled by the terebinths of Mamre, which are in Hebron, and built an altar there to the LORD. (Genesis 13:18)

It is in Hebron that an eternal covenant is decreed for the patriarch, whose name is then changed from Abram ("exalted father") to Abraham ("father of a multitude"). The voice of the divine declares:

> As for Me, behold, My covenant is with you, and you shall be the father of a multitude of nations. Neither shall your name any more be called Abram, but your name shall be Abraham; for I have made you the father of a multitude of nations. (Genesis 17:4-5)

It is in Hebron, "by the terebinths of Mamre," that three men, apparently an avatar for/of God, flanked by two angels, appear to Abraham:

> As he sat in the tent door in the heat of the day … he lifted up his eyes and looked, and, behold, three men stood opposite him; and when he saw them, he ran to meet them from the tent door, and bowed down to the earth… (Genesis 18: 1-2)

It is at Hebron that Abraham's God announces the destruction of the two prototypical cities of evil, Sodom and Gomorrah. Abraham's challenging question echoes through time: "Shall not the Judge of all the earth do justly?" (Genesis 18:25). Abraham intercedes on behalf of the inhabitants, asking if the cities might be spared for the presence of even ten righteous souls. In the end not even ten are worthy of being saved, and the two cities of iniquity are consigned to consumption by fire and brimstone.

It is at Hebron, also known biblically as Kiryat Arba, that Abraham's wife, Sarah, dies. Abraham declares to the people of the land:

> I am a stranger and a sojourner with you: give me a possession of a burying-place with you, that I may bury my dead out of my sight. (Genesis 23:4)

Abraham designates the place of her burial as a local grotto known as the Cave of Machpelah, and proceeds to buy it from Ephron the Hittite for "four hundred shekels of silver, current money with the merchant" (Genesis 23:16). To the Israelites this is nothing less than a contract, eternal and immutable. It is the moment in time at which the Jewish people, as descendants of the great patriarch, lay a legal claim to the cave and the land on which it is situated. It will of course take generations before the city becomes truly Israelite.

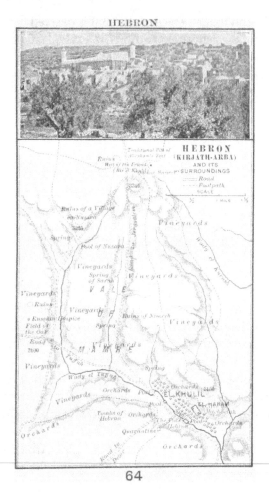

64

1899 Map of Hebron

After the great exodus from Egyptian bondage, Moses dispatches spies into the land of Canaan, who return from the Hebron area bearing a cluster of grapes of such size and weight that two men are required to carry it. The image is recalled to this day as the logo of Israel's Ministry of Tourism. Later, it is Joshua, who, having taken the reigns of power from Moses, also takes up the sword to lay waste to the city. Its king is slain "with the edge of the sword," along with all its inhabitants, leaving none alive.[5] The brutality is heinous, if the biblical record is to be believed, but also "par" for the ancient world's blood-drenched course. However, it will take another Israelite, whose cunning is matched by his charisma, to immortalize it as a city of kings:

> And David brought up the men who were with him, each with his household; and they dwelt in the cities of Hebron. And the men of Judah came, and there they anointed David king over the house of Judah. (2 Samuel 2:3-4)

Every king in ancient Israel would henceforth be "anointed," and made a "messiah" (from the Hebrew verb *mashakh/* "anoint"), when a priest or prophet would pour a horn of oil on the head of the person selected. It is in Hebron that the tradition begins, of another "anointed" king/ messiah, who will sit on the throne of David in the distant future. The attachment of the Jewish people to this place would henceforth become such that they would cling to it tenaciously in future generations.

After the Babylonian conquest of the southern kingdom of Judah in the sixth century before the Common Era, the city was emptied of inhabitants, remaining unoccupied for the next three hundred years. Nevertheless, a hilltop settlement was established one kilometer to the north of the ancient mound, or tel, where Hebron itself was presumably located. It corresponds well with the description by Josephus of a terebinth tree. Its Arabic name, Nimra, is most likely a corruption of the original, Abraham's Mamre.[6]

Babylonian rule was replaced by Persian suzerainty, and after Alexander the Great's conquest in 332 B.C.E., a new Greco-Hellenistic city was established to the east of the original site, nestled in a wadi. Various inscriptions on tombs in the vicinity have been identified, bearing variations of the god of the Edomites, Cos, and suggesting that the residents in those days were mostly of Edomite stock. The Edomites had drifted into the region following Jerusalem's fall in 586 B.C.E., and later became

---

5 See Joshua 10:37.

6 See Lukasz Niesiolowski-Spano, Jacek Laskowski, *The Origin Myths and Holy Places in the Old Testament: A Study of Aetiological Narratives* (London: Routledge, 2014), 115ff.

known as Idumeans. The region of Idumea would be forcibly converted to Judaism under the Jewish Hasmonean Dynasty, the descendants of the freedom fighters known as the Maccabees. In the bitterest of ironies, the grandiose structure over the ancient Cave of Machpelah would be built by one Idumean in particular, the notorious Herod the Great.

In any case, the Hasmoneans had a vested interest in sanctifying the site in their attempt to fuse religion with national restoration and awakening. An important Hebrew text dating from around 109 B.C.E., known as the Testaments of the Twelve Patriarchs, declares that Reuben was interred in the double cave at Machpelah, next to his ancestors.[7] The location of the cave was therefore well recognized by at least the early second century before the Common Era. Simultaneously, the Jewish character and spirit of Hebron strengthened and intensified. Nonetheless, the city's Jewish Renaissance was destined to fade. During the Great Revolt against Rome, from 66 to 70 C.E., Hebron's inhabitants struggled valiantly against the advancing legions, only to be crushed along with their compatriots across Judea.

Segregated Centuries

Many sorrowful centuries ensued, as Hebron's Jews experienced the suzerainty of multiple masters. Herod's massive shrine was most likely open to the sky, but the Byzantines built a church into its interior. Following the Arab conquest of the land in the year 638, a mosque replaced the church. Known today as the Ibrahimi Mosque, its four weathered gray domes are topped by golden crescents.

The positioning of the shrine itself was always a mystery, its access running north by northwest and east by southeast. This made it difficult to be used as a synagogue, since it did not face Jerusalem. As a mosque, it did not face Mecca, and as a church, it did not face the rising sun. It could only be assumed that the alignment of the fabled the caves beneath determined the direction of the structure above. As for the darkened hollows underneath the floor of the great monument, there swirled even more mystery. Due to the religious sensitivities surrounding the structure and the ground below, access to visitors was denied, and its subterranean chambers were shrouded in fanciful conjecture. Consequently, over the last two millennia, they lay beyond the reach of the most intrepid explorers, with only a few exceptions.

The medieval Arab geographer, Al-Maqdisi, wrote as follows in the

---

7 *Testaments of the Twelve Patriarchs,* 1, 7.

year 985:

> Habra (Hebron) is the village of Abraham al-Khalil (the Friend of God)... Within it is a strong fortress ... being of enormous squared stones. In the middle of this stands a dome of stone, built in Islamic times, over the sepulchre of Abraham. The tomb of Isaac lies forward, in the main building of the mosque, the tomb of Jacob to the rear; facing each prophet lies his wife. The enclosure has been converted into a mosque, and built around it are rest houses for the pilgrims, so that they adjoin the main edifice on all sides.[8]

Tomb of the Patriarchs, interior

The city changed hands again when it was captured by the Crusaders in the year 1100. With them came a cadre of monks, who, in 1119, claimed to have come upon a vertical indentation in the floor, from which they could feel a soft flurry of air. According to their account, they made an entrance into the depths of the structure, breaking through walls and finding, to their complete amazement, the skeletal remains of the biblical patriarchs still in place. That at least was their fantastical account. It was enough, however, to promote enormous interest in the site, and contributions sufficient to finance the Crusader church, which to this day occupies the inner parts of the walled enclosure. Inside the structure are six imposing sepulchral monuments, properly known as cenotaphs. Though certainly not the actual tombs of the three patriarchal couples, Abraham and Sarah, Isaac and Rebecca,

---

8   See Basil A. Collins, trans., *Best Divisions for Knowledge of the Regions: Ahsan al-Taqasim fi Marifat al-Aqalim* (Reading, England: Garnet Publishing, 2000).

and Jacob and Leah, they lend to the place a sense of solemnity and an almost otherworldly aura.[9]

Arab treatment of the Jews of the city had been relatively mild, but the Crusaders expelled them entirely. Nonetheless, the greatest Jewish philosopher in the world, Moses Maimonides, managed to visit the location in the year 1166, noting:

> On Sunday, 9 Marheshvan (17 October), I left Jerusalem for Hebron to kiss the tombs of my ancestors in the Cave. On that day, I stood in the cave and prayed, praise be to God, (in gratitude) for everything.[10]

In the year 1170, the great medieval Jewish traveler, Benjamin of Tudela, also managed to visit Hebron, and recorded his observations:

> Here there is the great church called St. Abram, and this was a Jewish place of worship at the time of the Mohammedan rule, but the Gentiles have erected there six tombs, respectively called those of Abraham and Sarah, Isaac and Rebekah, Jacob and Leah. The custodians tell the pilgrims that these are the tombs of the Patriarchs, for which information the pilgrims give them money. If a Jew comes, however, and gives a special reward, the custodian of the cave opens unto him a gate of iron, which was constructed by our forefathers, and then he is able to descend below by means of steps, holding a lighted candle in his hand. He then reaches a cave, in which nothing is to be found, and a cave beyond, which is likewise empty, but when he reaches the third cave behold there are six sepulchers, those of Abraham, Isaac and Jacob, respectively facing those of Sarah, Rebekah and Leah.[11]

The Kurdish Muslim conqueror, Saladin, retook the city in the year 1187, due in part to Jewish assistance in exchange for a letter of security. The Jews returned and continued to reside there until modern Israel's loss of the territory in 1948. Toward the end of the fifteenth century,

---

9 See C. Conder, PEQ (1881): 266-71; L. H. Vincent, E. J. H. Mackay, and F.-M. Abel, *Hebron: Le Haram El- KhaKl* (Paris: Editions Ernest Leroux, 1923); "Hebron," *EJ* 8:226-36; D. M. Jacobson, "The Plan of the Ancient Haram el-Khalil at Hebron," *PEQ* 113 (1981): 73-80; N. Millar, "Patriarchal Burial Site," *BAR* 11/3 (1985): 26-43.

10 Joel L Kraemer (2001). "The Life of Moses Ben Maimon," in Fine, Lawrence (ed.). *Judaism in Practice: From the Middle Ages Through the Early Modern Period* (Princeton, NJ: Princeton University Press), 422.

11 See Rich Cohen, *Israel Is Real: An Obsessive Quest to Understand the Jewish Nation and Its History* (New York: Picador, 2009), 193; Marcus Adler (ed.), Benjamin of Tuleda (Oxford: Henry Frowde, 1907), 25.

when the land was under Ottoman suzerainty, an additional testimony was recorded by the well-known Italian rabbi, Obadiah Bartenura, who was credited with reinvigorating the Jewish community of Jerusalem. He wrote:

> We reached Hebron, a small town on the slope of the mountain, called by the Turks Khalil. It is divided into two parts, one beside the Cave of the Patriarchs; the other opposite, a bow-shot farther away. I was in the Cave of Machpelah, over which the mosque has been built; and the Arabs hold the place in high honor. All the Kings of the Arabs come here to repeat their prayers... The Arabs remain above, and let down burning torches into it through a window, for they keep a light always burning there... Without, in the wall of the Cave, there is a small opening, said to have been made just after the burial of Abraham, and there the Jews are allowed to pray, but none may come within the walls of the cave, said to be the grave of Jesse, the father of David.[12]

From this it is evident that the historical exclusion of Jews from holy sites was a long established practice, and continued in Ottoman times. Yet, this is hardly a footnote in the tear-stained legacy of Jewish life under twenty centuries of foreign domination. The segregation of Jews had become a historical "given." In the early nineteenth century, an Egyptian Mamluk leader who led a rebellion against the Ottomans, Ali Bey al-Kabir, appears to have described each tomb as a separate room, guarded by iron gates and wooden doors with silver plating. He recorded:

> All the sepulchers of the patriarchs are covered with rich carpets of green silk, magnificently embroidered with gold; those of the wives are red, embroidered in like manner. The sultans of Constantinople furnish these carpets, which are renewed from time to time... I counted nine, one over the other, upon the sepulcher of Abraham.[13]

Not only did Jewish visitors continue to explore what they could of the patriarchal gravesites, but communities of Jews sprouted in the area, in spite of all the hardships. Between forty-five and sixty families of Spanish ("Sephardic") ancestry had arrived by the middle of the nineteenth century, along with fifty families of Polish and Russian extraction. Among those were members of the Hasidic movement known

---

12 Elkan Nathan Adler, ed., *Jewish Travelers* (London: Routledge, 1930, 2014), 233.
13 John Kitto, *The Illustrated Commentary on the Old and New Testaments* (London: Charles Knight & Co., 1840), 57.

as the Lubavitchers.[14] If continuity of residence means anything, the Jews made sure that their claim on the city was as authentic as anyone's.

Looking ahead to the twentieth century, the Grand Mufti and uncontested leader of the Palestinian Arabs during the British Mandate, Mohammed Amin al-Husseini, unleashed in 1929 a series of violent atrocities ("pogroms") aimed at Jewish communities in Jerusalem, the Galilean town of Safed, the Gaza Strip, and Hebron. Well over one hundred Jews were murdered, including families in their own homes, and hundreds more were wounded. In Hebron, the death toll was sixty-seven. The four hundred thirty-five surviving Jewish residents were then forced by the British to evacuate.[15] It was a familiar pattern, as Arabs designed, by fear and intimidation, to empty the land they called "Palestine" of Jewish presence. From Israel's War of Independence in 1948 (after which Jordan formally annexed East Jerusalem, Judea and Samaria) until the 1967 Six-Day War, Jews were unable to visit the city and its hallowed cave-shrine at all.

Raiders of the Lost Tombs

When modern Israel emerged victorious after its rout of Arab armies, Jews found that they had access to the tombs for the first time in seven hundred years.[16] Did Israel re-expropriate the site, ejecting the Muslims and closing their mosque? On the contrary, under Israeli control, people of all faiths were given free and open access to this and all holy places in the newly conquered land. It should surprise no one that in the wake of two decades of ad hoc annexation by Jordan, Jewish settlers should again call Hebron their home and announce their intention to stay. Nor should it come as a surprise that so much of their attention should be focused on the great Herodian structure, parts of which were reconstructed to include two small synagogues, and its subterranean parts, including the cave of Machpelah itself.

Not long after the Six-Day War, then Defense Secretary Moshe Dyan, Israel's dashing military genius and erstwhile archaeologist, known for the patch he wore over his left eye, opened a blocked off aperture on the far end of the building and lowered a twelve-year-old Israeli girl named Michal, whose narrow torso was sufficiently slender, into the shaft that

---

14 See John Wilson, *The Lands of the Bible Visited and Described*, Volume 1 (Edinburgh: William Whyte, 1867), 357ff.; Martin Sicker, *Reshaping Palestine: From Muhammad Ali to the British Mandate, 1831-1922* (Westport, CT: Praeger, 1999), 33ff.

15 See Glick, op. cit., 29.

16 See Hassner, op. cit., 73.

was said to lead to the burial caves themselves. Hauling her back up after her adventure, Michal related seeing a diminutive chamber in the depths below, along with a stone staircase ascending upward toward the main floor. She produced a crude drawing of what she had seen, enough to whet the appetites of many a frustrated explorer over the decades to come. Israel made no more official attempts to peruse the hidden cavities of shrine, though, as one Hebron resident recalled, "We had many Jewish visitors coming to the city and there was not a tense relationship with the settlers and the Palestinian people."[17]

Michal, lowered into shaft leading to burial caves

In 1982, the Jewish community spokesman of Hebron, Noam Arnon, and five accomplices, inspired by the map drawn by twelve-year-old Michal, made an "unofficial" foray to pry open the stone which sealed the burial chambers of the sacred shrine, whereupon they ventured downward. Such actions, considered illegal by the Muslim authorities, might have resulted in criminal charges and imprisonment. Arnon nonetheless declared:

At the time, we were beginning to hear opinions in politics and even

---

17 Tovah Lazaroff, "The Goldstein Massacre's Shadow on Hebron," *Jerusalem Post*, Mar. 21, 2019: https://www.jpost.com/Arab-Israeli-Conflict/The-Goldstein-massacres-shadow-on-Hebron-584234.

academia that the entire story of the Machpelah and Abraham was a myth and not based on fact. We needed to prove that the Biblical account was accurate.[18]

The adventure was orchestrated for the Hebrew month of Elul, just prior to the highest holy day, Yom Kippur, when entreaties for divine absolution are offered and the din of prayer mingled with the blowing of rams' horns (*shofarot*) is almost deafening. Ample distraction being present, the team made their way to the exact spot where little Michal had descended, a decade and a half earlier. Arnon recalled:

After some searching, we found the opening. It was covered by a flag-stone in the middle of the floor in the public area. We had brought in some tools and at a point when the shofarot were the loudest, we pried the stone open.[19]

Tomb of the Patriarchs, synagogue

18 See Adam Eliyaho Berkowitz, "The Secret Chambers Hidden in the Cave of Patriarchs Will Astound You," *Breaking Israel News*, Jan. 11, 2017: https://www.breakingisraelnews.com/81859/secret-hidden-chambers-hebrons-cave-patriarchs-finally-revealed-photos/.
19 Ibid.

There before their eyes was the very staircase, chiseled into the bedrock, depicted by Michal. Arnon went on:

> We descended to the bottom where there was a long and narrow tunnel, sloping even further down, that opened into a chamber with a roughly tiled floor. I felt a slight wind and realized there had to be another opening. We followed the airflow and and tracked it to an opening in the floor covered by a stone. We pried it open and crawled down into a cave carved into the bedrock. In the back of the cave was another opening into a second cave with a low roof. At one point, I found myself crawling among human bones.[20]

Human bones? Archaeologically, no serious case could be made that these were patriarchal or matriarchal bones, but their presence was certainly enough to send chills up the spines of serious researchers. However, at the moment that the subterranean sleuthing of Arnon and his friends was becoming most compelling, the boisterous praying up above was winding down, and the accomplices assigned as lookouts beckoned them to surface and make an expedited egress, so as to avoid detection and retribution. While the escape succeeded, undetected, the Muslim Arabs discovered, on the following morning, that someone had descended into the forbidden recesses. The *Waqf* issued a formal complaint to the Jewish state, and the city's Arabs staged a major riot.

In any case, Arnon had been able to retrieve a selection of broken potshards from his underground foray, and, submitting them for testing, found that they dated to the period of the First Temple – the dynasty of David, Solomon and their descendants. The Israel Antiquities Authority went further, determining that the caves could safely be dated to the Bronze Age, placing them firmly at a time concurrent with the days of the patriarchs and matriarchs. Arnon declared, "All of the archaeological evidence shows the site is historically Jewish, dating back three thousand years." Such evidence is, however, quite irrelevant to UNESCO, whose resolutions have attempted to sever the Jewish connection with the site. Said Arnon:

> I did carbon-14 dating on the structure itself and it dates back to before the first century C.E. The Muslims can't cope with the facts and the actual history of the site, so they don't even make an attempt. They simply object to Jews entering the site, as they did throughout the Ottoman Empire. When the United Nations supported the claim, it was not

---

20 Ibid.

based on fact or any semblance of truth. It was pure anti-Semitism.[21]

Unfortunately, Jewish claims on Hebron would be seriously under-mined by one of the most appalling episodes in the legacy of modern Israel.

Modern Madness

Tragically, Israel's policy of moderation and tolerance was shattered when, on February 25, 1994, around five o'clock in the morning, an American Jewish settler and well-known doctor, who lived in the nearby settlement of Kiryat Arba, entered the mosque in the Tomb of the Pa-triarchs. On the Jewish calendar it was Purim, the feast of Esther, when, in ancient times, the Jews of Persia, faced with a murderous plot against them, rose up and slaughtered their enemies. Brandishing an automatic rifle, Baruch Goldstein opened fire on a group of Muslim worshippers, unleashing horrible carnage and killing thirty. An "ear-witness" to the horror recounted, "I woke up to hear people shouting and the mosque [loudspeaker] calling for people to go to the hospital to give blood."[22]

Goldstein was beaten to death by those in the vicinity, and Israeli Prime Minister Yitzhak Rabin denounced him as a "degenerate mur-derer" as well as "a shame on Zionism and an embarrassment to Juda-ism."[23] Many years later, the same Noam Arnon who had clandestinely explored the Caves of the Patriarchs commented: "No one is trying to justify or downplay the Goldstein massacre;" but he added that nothing remotely resembling such violence, perpetrated by Jews, has occurred since.[24] Nonetheless, the damage done to the dream of peaceful rela-tions was irreversible. The massacre transpired during a series of events that were destined to create a new and deadly status quo. The terrorist organization Hamas responded with multiple murders and bombings. Following the Oslo "Peace" Accords of 1993, a wave of terrorist attacks against Jews was unleashed, continuing to the present. In 2002, twelve Jews were indiscriminately attacked and killed in Hebron, on Worship-pers Way, linking the city with Kiryat Arba, the urban Israeli settlement on its outskirts.

---

21 Ibid.

22 Lazaroff, op. cit.

23 Clyde Haberman, "West Bank Massacre: The Overview; Rabin Urges the Palestin-ians To Put Aside Anger and Talk," *New York Times*, Mar. 1, 1994.

24 Lazaroff, op. cit.

Hebron, with Tomb of the Patriarchs

In response, Hebron was divided in 1997, with Israel ceding eighty percent of the city to the Palestinian Authority, simultaneously prohibiting her own citizens from so much from entering that sector. Jewish population growth in Hebron, consisting of just eighty families, has stalled, as Jews are restricted in their movement to a mere three percent of the city. The 220,000 Arabs who still reside there are outside of Israeli control and Israeli law. Extending Israeli sovereignty to the whole of Hebron is something supported by the majority of Israel's Likud Party, but until such dreams are realized, it will remain as one more Berlin-like, divided city in a troubled region, seething with unrest.[25] Ironically, the creation of a Palestinian state will not eliminate walls, separation or segregation; it will only intensify, solidify and make permanent the sad reality. If the silent caves underground have anything to say, it is that this has always been a Jewish land, and dividing it again, to accommodate the Palestinian Authority or a Palestinian state, is worse than folly. It is madness.

The City of Abraham

On an embankment to the west of the old city of Hebron lies an archaeological site known as Tel Rumeida, an Arabic word meaning "ash

---

25 Ibid.

gray," in reference to a fire that consumed the area. It is believed by a number of Jewish archaeologists to be the original site of biblical Hebron. Excavations, under Professor Philip C. Hammond of the American Expedition to Hebron, were conducted between 1963 and 1966. An additional dig in 1999 was headed by Emanuel Eisenberg, of the Israeli Antiquities Authority. Eisenberg returned in 2014 with Professor Shlomo Ben-David of Ariel University, to continue the work of uncovering what proved to be an impressive archaeological site.[26]

Today, an ancient wall, three meters high, is easily discernible, and is imagined, some four thousand five hundred years ago, to have reached a height of ten meters. Another, opposing wall, dates to around 1,700 B.C.E., the general time frame of the biblical Abraham. Between the two lies the remains of an ancient staircase, over four thousand years old. It is speculated that if the excavations could be continued, the ancient gates of the city would be revealed.

Site of ancient Hebron

Might this be the spot where Abraham purchased the burial cave for Sarah? Such a discovery would, however, require digging up the current city street, which, due to political and cultural sensitivities in the con-

---

26 https://www.hebronfund.org/tel-hebron-archaeological-park-and-admot-ishay-tel-rumeida-neighborhood/.

temporary tinderbox, has not been attempted. Remains of ancient Israelite four-chambered homes at the site date from between the Bronze Age and Iron Age, when King Hezekiah sat on David's throne in Jerusalem. As excavations progressed, an ancient granary and wine press were uncovered as well. Also from the First Temple period were workshops, kilns, and humongous vessels for storing oil and wine. There were other structures from the early Roman era, along with pottery, coins and jewelry.[27]

A seal impression discovered in one of the houses bears the Hebrew inscription "To the King," and beneath it, the word "Hebron." An "unconsidered trifle" such as this amounted to a proverbial "smoking gun," and another of the many facts underground, beckoning Jewish settlers to return.

Beit Menachem, above archaeological site of ancient Hebron

Clearly, Jews lived here from distant antiquity, though they laid legal claim to the land once more when, in 1807, Sephardic Jews (whose roots went back to medieval Spain and the Arab world) purchased the entire

27 Tovah Lazaroff, "From Bronze Age to First Temple: Archaeological Site Set to Open in Hebron," *Jerusalem Post*, Oct. 17, 2018: https://www.jpost.com/Israel-News/From-Bronze-Age-to-First-Temple-Archaeological-site-set-to-open-in-Hebron-569506.

area around Tel Rumeida, transferring the ownership of a portion of it to Ashkenazi (European) Jews. In 1811 an Egyptian Jew, Rabbi Haim Yeshua HaMitzri, secured another land purchase, encompassing, in addition to the tel, the tombs in the vicinity associated with Ruth and Jesse, the father of David. A nearby structure dates back fifteen hundred years and may have been used as a synagogue. To this day, the tombs contain a small synagogue and study hall. There were serious plans to establish a Jewish residential quarter near Tel Rumeida, but Arab unrest associated with the anti-Zionist riots of 1929 instead resulted in the expulsion of the very Jews who had bought the land and hoped to build on it.[28]

Only in 1984 did the government of Israel grant permission to its own citizens to construct a residential quarter at this location. Seven Jewish families moved into seven mobile homes, struggling through hardship and deprivation to be in the vanguard of a reborn Jewish presence in Hebron. After the Wye Accords of 1997 divided the city into two sectors, the situation only deteriorated. In August of 1998 any semblance of peaceful coexistence that yet prevailed was shattered, when a Palestinian Arab infiltrated the fledgling settlement and murdered Rabbi Shlomo Ra'anan, whose grandfather was the renowned Rabbi A. I. Kook.[29] Responding to the settlers' vulnerability, the state of Israel approved a permanent new residential building for the community, which came to be called Beit Menachem, in honor of the legendary leader of the Chabad Lubavitch movement, the late Menachem Mendel Schneerson. It is beneath this new construction that the archaeological site lay in protected solitude.

Not surprisingly, with the opening of an archaeological park to the public, showcasing the many finds of Tel Rumeida, voices of angry discord, emanating from both the international and Israeli left, have been broadcast loudly. "A new Israeli archaeological site is tightening the noose around suffering Hebron Palestinians," wrote Jerusalem-based freelance journalist Ben Lynfield.[30] Emek Shaveh predictably chided the sponsors of the park for utilizing archaeology to advance a political

28 https://www.hebronfund.org/tel-hebron-archaeological-park-and-admot-ishay-tel-rumeida-neighborhood/.

29 See Sara Bedein, "Chaya Raanan: Profile of Courage in Hebron," *Israel Behind the News* Sep. 18, 2002: https://israelbehindthenews.com/chaya-raanan-profile-of-courage-in-hebron/3802/.

30 Ben Lynfield, "New Israeli Archaeological Site Tightens Noose Around Suffering Hebron Palestinians," *N World*, Oct. 23, 2018: https://www.thenational.ae/world/mena/new-israeli-archaeological-site-tightens-noose-around-suffering-hebron-palestinians-1.783465.

agenda:

> After decades of attempts to build and live in Tel Rumeida, it appears that the settlers have realized that by using tourism and archaeology, they can conquer the tel without having to admit that it is a settlement. In Hebron, the struggle over heritage cannot be separated from the struggle over land. Archaeology, in this context, has become an effective tool for the settlers in their efforts to entrench their hold over Hebron.[31]

The liberal advocacy and activist group in Israel, Peace Now, chimed in as well:

> The archeological site in Hebron is a settlement in every respect, as it is intended to be. Archeology serves as a tool for the government to solidify its control over the territory as a future sovereignty claim. It is meant to change the public domain in Hebron and to influence public consciousness in order to strengthen the settlement enterprise... The establishment of the archaeological park as a tourist attraction in Tel Rumeida is intended to bring thousands of visitors deeper into the Palestinian area of Hebron, and to turn additional parts of Hebron into an Israelis-only area.[32]

This, then, is the "politically correct" view of Jewish settlers on historically Jewish land. It strikes some as odd that Jewish archaeology is singled out for such criticism, when archaeology on behalf of any other political or cultural "agenda" may proceed unscathed. An example is Native American archaeology, which, it is argued by some, can be used to further the cause of tribal populations in the United States. One professor of anthropology at Fort Lewis College in Durango, Colorado, sees it as a tool in "Confronting Cultural Imperialism." He wrote: "The ethical, legal, and research-oriented tools of archaeology can encourage Native American self-determination..."[33] Does this not sound like an "agenda?"

It bears repeating that Jews (unlike Spanish conquistadors and the host of European colonialists who followed) have never, in their long, bloodstained history, attempted to colonize or expropriate the territory of any other people. The only claim Jews make or have ever made is to the territory promised to them and ratified with the words of a di-

---

31 https://alt-arch.org/en/tel-rumeida-hebrons-archaeological-park/.

32 https://peacenow.org.il/en/new-touristic-settlement-opens-tel-rumeida-hebron.

33 https://www.sapiens.org/archaeology/native-american-archaeology/.

vine covenant between the God of Israel and the "father of a multitude," Abraham: "Unto your seed will I give this land" (Genesis 12:7). Some promises are transient, some insincere, others simply false. Some, however, are inviolate.

Of course, the politically correct left will ask what is to become of the majority Arab population of Hebron. The answer lies in the fact that there are really only two doors to the future. Behind one is a newly minted Palestinian state, which experience tells us would look something quite similar to what already exists in Gaza. Indeed, surveys indicate that if the Palestinians living in Judea and Samaria had their say, they would oust the current leadership of the Palestinian Authority and substitute a Hamas-led government. Fantasize all you want about two states living side-by-side in peace, but the harsh reality is otherwise. We hardly need reminding that the Middle East is not Switzerland, divided among German and French-speaking populations, where men wearing lederhosen come yodeling over the hillsides. At the risk of igniting a firestorm among left-wing critics, we might well observe that the so-called "two-state solution" amounts in reality to the "final solution."

Door Number One presents us with a terrorist state, consumed with blind hatred for Israel and the Jewish people. Wishful thinking cannot change the fact that there is no such thing as a "civil society" in today's Palestine. Its territory will be *Judenrein*, and without the Israel Defense Force to protect them, the Jewish settlers who live in Hebron and in other places across the ancient Jewish heartland will either be forced to evacuate, as homeless refugees, or face certain slaughter. Rest assured, the leftist intelligentsia will raise no outcry, claiming instead that Jews should never have moved into such places to begin with. Never mind that they are literally living on top of the Bible.

Door Number Two presents the only other realistic possibility. It involves the extension of Israeli law and sovereignty over the territories, won in a defensive war, and which reveal evidence of Jewish habitation for at least the last three thousand years. Far from being an "apartheid" portal, Door Number Two presents a future in which the Palestinian Arab population would be offered full Israeli citizenship to match the nearly two million Israeli Arabs who already enjoy full rights as citizens of the Jewish state. Of course, no one would be forced to take such citizenship. The Arabs who so choose could continue to live as they are or find residence in the scores of other Arab countries in the world today. Aside from continuing the current unhappy status quo, there are no other options, except, perhaps, a Door Number Three: war. The most important reality to consider is that whichever door is opened, consequences will certainly follow, for the region and for the world. The choice, unclouded by the din of political correctness, should be clear enough.

CHAPTER SIX

OWNING HISTORY – ANCIENT JERICHO AND THE 9,000-YEAR-OLD MYTH

"**P**ALESTINE," wrote Mark Twain, "sits in sackcloth and ashes:"

Over it broods the spell of a curse that has withered its fields and fettered its energies. Where Sodom and Gomorrah reared their domes and towers, that solemn sea now floods the plain, in whose bitter waters no living thing exists – over whose waveless surface the blistering air hangs motionless and dead – about whose borders nothing grows but weeds, and scattering tufts of cane, and that treacherous fruit that promises refreshment to parching lips, but turns to ashes at the touch. Nazareth is forlorn; about that ford of Jordan where the hosts of Israel entered the Promised Land with songs of rejoicing, one finds only a squalid camp of fantastic Bedouins of the desert; Jericho the accursed, lies a moldering ruin, to-day, even as Joshua's miracle left it more than three thousand years ago...

Jericho, designated in pagan lore as the "City of the Moon" (where the lunar orb was doubtless venerated in antiquity) is today nothing more than an excavated mound, silhouetted against the relentlessly brilliant sky. For the better part of the last three millennia, it has lain in long melancholy repose. The incessant, blazing heat of the Judean Desert bleaches away all colors, leaving only a chalky gray, punctuated by a hearty grove of assorted palm trees, where a subterranean stream makes its abrupt egress. A green patch of fertility stands out against the harsh tones of the otherwise desolate landscape. In this barren land visual contrast is blurred, and distinct shapes are all but obliterated, with only light and dark patches of pale pastel interrupting the trackless, arid wilderness round about, stretching in all directions. The fact that a great and mighty city stood here, thousands of years ago, would be difficult for the rational mind to fathom, were it not for the presence of a huge, layered, artificial heap of stone and rubble. Its once-smooth outlines are now interrupted by a series of great gashes, carved into the remains of the ancient site by generations of obdurate excavators, intent on determining either the veracity or mythology of Joshua's conquest.

Ancient Jericho - Tell es-Sultan

Lamentation, Celebration and Song

Jericho has the considerable distinction of being the world's oldest city as well as its geographically lowest, being some two hundred fifty meters below sea level. There is archaeological evidence of human habitation here, dating back more than eleven thousand years, made possible by the fact that it is the only oasis in the lower Jordan Valley. However, beyond the science of archaeology, Jericho is a totem in today's Israel, representing the unshakable conviction that this is the eternal homeland of the Jewish people. It is nonetheless a homeland rife with controversy, dispute and violence. From the birth of the Jewish state in 1948 until the Six-Day War of 1967, the entire territory from East Jerusalem to the Jordan River was devoid of Jewish presence, having been annexed by the Hashemite kingdom of Jordan.

While Israel had won its independence as modern nation, it lacked a great deal in terms of territorial integrity, and while it was able to secure a fragile armistice with its hostile neighbors, there was never a genuine peace. Jerusalem was a divided city, the Berlin of the Middle East. The east side was in Arab hands, the west side was in Jewish hands, and coursing through its midsection was a perilous mine field, cordoned off with barbed wire. A sense of foreboding pervaded the atmosphere on both sides, and for the city's Jewish inhabitants there were constant reminders of the many places that were strictly off-limits.

In the spring of 1967 Israel's most celebrated songstress, Naomi Shemer, composed her beloved melodic masterpiece, "Jerusalem of Gold." Commissioned by Jerusalem's illustrious mayor, Teddy Kollek, it was presented in honor of the Israeli Song Festival, being held on May 15, the night after the country's nineteenth Independence Day. Its melancholy verses are still intoned today, its chorus reading (in English translation): "Jerusalem of gold, and of copper and of light; for all of your songs am I not a lyre?" One of the song's verses is particularly significant, given the time it was written: "We cannot go down to the Dead Sea by way of Jericho…" Indeed, Israeli Jews were not able to visit the city made famous by Joshua, or venture to any other destination in Judea and Samaria, the biblical heartland.

The events of June of that year, however, were destined to change everything. In six days, Judea and Samaria, the so-called West Bank of the Jordan River, fell to Israel. The reality of that moment was almost beyond belief, and the lament Naomi Shemer had just completed suddenly seemed out of place and inappropriate. Hastily, she composed a final verse, adding the words (in English translation): "And again we go down to the Dead Sea by way of Jericho." Shemer's mournful melody had become a song of celebration. While none of the newly conquered territories were formally annexed, Israelis were now able, for the first time since the Jewish state was born, to visit the place where Joshua's invasion began. Might it even be possible for Jews to live in the vicinity of the storied city whose walls tumbled before the Israelites?

Owners of History?

In the interest of historical fairness, however, it is not out of place to ask a pointed question. How far back can one trace the Jewish claim on this city, and what are the implications for both Israelis and Palestinian Arabs who claim it today? In the second half of the twentieth century, archaeology and geopolitics were destined to collide. In May 2011, Palestinian President Mahmoud Abbas made an audacious assertion:

We say to him [Netanyahu], when he claims – that they [Jews] have a historical right dating back to 3,000 years B.C.E. – we say that the nation of Palestine upon the land of Canaan had a 7,000 year history B.C.E. This is the truth, which must be understood, and we have to note it, in order to say: "Netanyahu, you are incidental in history. We are the people of history. We are the owners of history."[1]

This is most curious, in light of the Palestinian claim that archaeology should never be used for political ends. To be sure, all that is required is a modicum of good archaeology to reveal the unvarnished truth about who laid claim to this land and when. The one bit of certainty is that the Canaanites and neolithic folk who inhabited this incredibly ancient city bore no relation whatever to today's Palestinian Arabs. Moreover, these proud ancient pagans, should they somehow be brought back from dusty death and interviewed, would be in deep shock at the very suggestion that they might be forced into the mold of "honorary Muslims." The general outlines of history are clear enough, and the Palestinians are by no means its "owners."

It should in any case be asked, regarding Abbas' 7,000 B.C.E. claim: to what was he referring? Clearly, his reference was to the city of Jericho, where we find a pre-Canaanite neolithic tower dating back some nine thousand years, before "civilization" as we know it even existed. An Al-Jazeera-produced documentary, entitled "Looting the Holy Land," made even more bombastic assertions, loudly trumpeting Jericho as a treasure of the Palestinian people, dating back ten thousand years. What we see is that archaeology has become one of the "front line" items in the ongoing propaganda campaign against the Jewish state.

Given such bombast, it is hardly surprising that in today's Middle East, Jericho has continued to occupy headlines. Of course, Jews did not suddenly arrive in Jericho after the Six-Day War. Notably, a fifteen-hundred-year-old synagogue was uncovered during excavations conducted in 1936 under the auspices of the British Mandate's Department of Antiquities. Aside from the ancient tel itself, the remains of the archaeologically important Shalom Al Yisrael Synagogue bear witness to a Jewish presence in the city going back at least to the sixth or seventh century C.E. Its mosaic floor features exquisite depictions of the Ark of the Covenant, the menorah, the *shofar* and *lulav* (the frond of the date palm

1 "Mahmoud Abbas," Official Palestinian Authority TV, May 14, 2011; Itamar Marcus and Nan Jacques Zilberdik, "Abbas to Netanyahu: 'You are incidental in history'," *Palestinian Media Watch*, May 24, 2011: https://palwatch.org/page/2800.

used during the Feast of Tabernacles), and contains the words "Shalom Al Yisrael" ("Peace on Israel"), after which the site was named. A private home was subsequently built over it, whereupon the local Arab family who owned it took to charging an entrance fee to see the intricately designed floor.[2]

Shalom Al Yisrael Synagogue

In the wake of their victory in June 1967, the Israelis extended military supervision over the site, while the Arab family was granted administrative control. Jewish tourists began frequenting the house in order to pray and of course to glimpse this archaeological link with their ancient homeland. To Israeli authorities it seemed more than odd that such a Jewish treasure should be overseen by Arabs, who had, after all, built their house after the synagogue had been discovered. One need only imagine the outcry if Jews had constructed a home over a historically important Islamic site. This is of course the classic double standard of today's Middle East.

In 1987, the Israelis determined that the time had come for Jews to take charge of what was obviously a Jewish site. Fair compensation was offered to the owners, as a kind of "eminent domain" deal, but it was unceremoniously refused. At that point, and only at that point, the mosaic floor, the house which stood above it, and a small section of an

---

2 See Judy Lash Balint, "The Lost Jewish Presence in Jericho," *Jerusalem Post*, Jan. 21, 2012: https://www.jpost.com/features/in-thespotlight/the-lost-jewish-presence-in-jericho.

adjacent farm, was taken over on behalf of the state of Israel. The Israelis did not, however, hold the property in perpetuity. Ultimately, the site was transferred to the Palestinian Authority, under the provisions of the Oslo Accords of 1993, with the understanding that it would be guarded by an official security detachment. Open access to the synagogue was also to be guaranteed for those wishing to visit.

Nevertheless, at the beginning of the al-Aqsa Intifada, on the night of September 28, 2000, vandals, in an act of desecration, entered and set fire to the site, damaging the house above in the process. Israeli fire trucks were prevented by Arab vandals from dousing the flames.[3] The Torah scroll stored at the synagogue miraculously survived. The Municipality of Jericho repaired the damage, but the Torah was removed to the nearby Israeli settlement of Mevo'ot Yericho. Not until 2005 was a group of Israelis able to revisit the synagogue. The Israel Defense Force thereafter decided to allow visits only on the first day of every Jewish month, for the sole purpose of conducting prayer services, though by 2007 weekly visits became possible.[4] In the near term, the synagogue appears safe. It is fair to ask, however, what will happen if and when the I.D.F. withdraws entirely and a Palestinian state is born. How many Jewish holy places may be expected to survive?

Digging for the Truth

In response to Abbas' claim it is also important to consider, not just the Canaanites, but the significant Jewish habitation at Jericho for thousands of years. Whether Joshua's tumbled wall may be found is, however, another question, requiring at least a modicum of background in the excavation of the site. In 1930 a professor at England's Liverpool university, John Garstang, brought a team to British-controlled Palestine, hoping to shed new light on the ruins of a place called Tell El Sultan – biblical Jericho. Garstang, born to Dr. Walter Garstang of Blackburn, was a man of impressive credentials. He was educated at Queen Elizabeth's, Blackburn, and Jesus College, Oxford. Following undergraduate studies in mathematics at Oxford, he turned his attention to archaeology. From 1930 to 1936 he excavated in Transjordan, at the most famous site of all, Jericho (Tell es-Sultan).[5]

---

3 Steven Carol, *Understanding the Volatile and Dangerous Middle East: A Comprehensive Analysis* (Bloomington, IN: iUniverse, 2015).
4 https://sites.google.com/site/mitzpeyericho/shalom-al-yisrael.
5 See John Garstang, *The Foundations of Bible History: Joshua, Judges* (Grand Rapids: Kregel, 1978).

John Garstang, July 1956

His methods, however, were not without serious deficits, the publication of his excavations being quite selective. Many finds were never published, and his maps and plans are often much too small, with important details missing. Consequently, modern researchers have criticized him for his apparent lack of care. Nonetheless, his excavations seemed to support his opinion that the biblical account is uncannily accurate. He claimed to have found evidence of catastrophic destruction and burning, including a collapsed wall, which he linked to Joshua's invasion. He dated the conquest to around 1400 B.C.E., perfectly in line with the traditional biblical chronology that places the exodus in the mid 1400s. In any case, his conclusions would by no means go unchallenged.

Adding an enormous layer of controversy was Garstang's fellow British archaeologist Dame Kathleen Kenyon, who came to Jericho in 1952. At the time, she was hoping to verify Garstang's findings with new technology, such as the experimental radiocarbon dating system. While she did find evidence of the same horrific destruction, the date assigned to it came as a shock to almost everyone – 1550 B.C.E. – a full century and a half earlier than Garstang's date. In fact, she concluded that Jericho was completely uninhabited at 1400 B.C.E., the days of the biblical chronology of Joshua. She commented:

> The archaeological remains indicated that many of the cities said to have been conquered by Joshua were not in existence at the end of the

late Bronze Age; nor was there any major change in material culture. Were it not for the Old Testament, no invasion would have been suspected at this time.[6]

This was exactly the sort of conclusion that would, over time, motivate contemporary "Israel-denying" Palestinians to declare that there never was a historical Jewish presence in this land.

Kathleen Kenyon

Being a devout Christian, however, Kenyon did not maintain that the biblical account of Jericho's destruction was fabricated out of whole cloth. Regarding the fabled walls, Kenyon, like Garstang, noted that the entire Jordan River Rift Valley is seismic, and that earthquake activity is not uncommon. She suggested that the walls could have collapsed due to an earthquake. She declared: "It would have been very natural for the Israelites to have regarded such a visitation as divine intervention…"[7] She always believed that archaeology was needed to prove the historic-

6 Kathleen M. Kenyon and Peter R. S. Mooney, *The Bible and Recent Archaeology* (London: British Museum Publications, 1987), 77.
7 Kathleen M. Kenyon, *Digging Up Jericho* (London: Ernest Benn, 1957), 262.

ity of the Bible, and more importantly, that archaeology was needed to enable the interpretation of what she called the "older parts of the Old Testament, which from the nature of their sources ... cannot be read as a straightforward record."[8]

On the other hand, many scholars were much more inclined to dismiss the account of the fall of Jericho as ancient Israelite folklore. The term is "etiology" – noticing a long abandoned ruin and then contriving a story to explain it. Nevertheless, advocates of the Bible as genuine history were by no means ready to surrender. For example, contemporary archaeologist, Bryant Wood, noticed that Kenyon had not actually examined Canaanite pottery found at the site. Instead, she based her conclusions on the absence of imported pottery that might have been expected in the Late Bronze Age. Garstang's pottery, by contrast, did in fact date to the Late Bronze Age, specifically to the key date of 1400 B.C.E.[9]

What's in a Pot?

When it comes to archaeology, the most underappreciated truism is that the pots hold the keys, and it is in them that the clues to almost every biblical mystery reside. To examine just a single broken shard is to know its origin, when and where it was fashioned, and by whom. The archaeological sleuth must first understand that by the Middle Bronze Age, pottery was already an industry produced in specialized workshops. The pottery fashioned in this region, known as the southern Levant, was of many shapes and sizes. Enormous storage jars were created to carry grain. Wine and olive oil were stored and shipped in smaller earthenware storage jars. Middle Bronze pottery vessels continued to be produced into the Late Bronze Age, changing in shape slowly. Mycenean imports became commonplace by Late Bronze II and were imitated by local potters from then on. The nature of imported goods in Aegean vessels is still open to discussion, and some scholars have speculated that diluted opium in particular was imported in small juglets.

Importantly, it was the lack of such imports on which Kathleen Kenyon based her conclusions, and that may have been her greatest mistake as an archaeological detective. With regard to Jericho, there is some additional evidence that should be considered. One detail of the previous

8 Ibid., 266.
9 See Walter C. Kaiser, *A History of Israel: From the Bronze Age through the Jewish Wars* (Nashville: Broadman & Holman, 1998), 152; William H. Stiebing, *Out of the Desert?* (Amherst, NY: Prometheus, 1989), 83.

excavations of Garstang and Kenyon involved the discovery of storage jars full of grain, clear evidence that the siege had occurred soon after the Spring harvest. According to the biblical account, an inhabitant of the city named Rahab sheltered two Israelite spies on her roof, under the flax, which had been left there to dry in the wake of the harvest. Such telltale evidence might suggest that Joshua's famed invasion might indeed have taken place in the spring, as the Bible relates. Moreover, since the jars were full, it is clear that the siege was very short, consistent with the biblical account of just seven days.[10]

Wood also observed that one area of the wall, which, unlike Kenyon, he dated to around 1400 B.C.E., had not collapsed. This is again consistent with the biblical record, indicating that the house of Rahab, literally built into the casemate wall, was spared by the Israelites. Furthermore, the Bible records that after its conquest, the entire city was burned. While Kenyon dated the destruction 150 years earlier, she nonetheless pointed out the remains of the fire: "This red and white streak is the remains of a tremendous fire so fierce that the ten-foot-thick walls were burnt through and through."

However, just when the tide might be turning in favor of the biblical account, new data came in. The actual remains of the grain in the jars were carbon dated. The results indicated not 1400 B.C.E. but indeed 1550 B.C.E. – just as Kenyon had asserted.[11] In fairness, the test contained an error factor of up to 110 years, but even that would be insufficient to align the grain with the biblical chronology of Joshua's conquest. Nevertheless, carbon dating has been shown to be less than accurate on multiple occasions, and the dispute might be reduced to whether one places greater trust in carbon-14 or in the physical evaluation of pottery remains. In the final analysis, the archaeological battle for Jericho is by no means over.

Habitation and Re-inhabitation

As tempting as it is among minimalists to dismiss out of hand the prominent place of Jericho among the conquests of Joshua, the fact that the city became an important habitation of later Israelites, over multiple centuries, is well attested by a preponderance of facts underground. According to biblical law, the city was cursed following the Israelite con-

---

10   Bryant Wood, "Did the Israelites Conquer Jericho: A New Look at the Archaeological Evidence," *BAR* 16/2 (1990): 44-59.

11   See Piotr Bienkowski, "Jericho Was Destroyed in the Middle Bronze Age, Not the Late Bronze Age," *BAR* 16/5 (1990): 45, 46, 69.

quest and left abandoned and unoccupied. However, the sullen vestiges of a characteristically Israelite four-room house, perched on its eastern slope and dating to the tenth and ninth centuries B.C.E., bear witness to the city's renewed habitation during the Judean monarchy. Later remains, from the 600s B.C.E., testify to the growth of the city into a major settlement, only to be destroyed within a single century by the great Babylonian conquest, which brought a final and devastating end to the kingdom of Judah. The city later became Persian, and later still a private estate of Alexander the Great.

The Jewish people, having returned from their exile, nonetheless clung tenaciously to the land, fomenting a great revolt against Syrian occupation in the 160s B.C.E., under a family of freedom fighters known as the Maccabees. Jericho again became Jewish, attested by the remains of an aqueduct dating from the days of the dynasty they founded, the Hasmoneans. While Tell es-Sultan was by then a covered ruin, several low mounds nestled along the banks of an intermittent stream bed known as the Wadi Qelt represent the Hasmonean settlement and palace of that period.[12]

Herod's Winter Palace at Jericho

The next major conquest was that of the Romans. It was the illustrious Mark Anthony who made a gift of Jericho to his Egyptian lover, Cleopatra, and it was the notorious King Herod the Great who leased

12 See Jerome Murphy-O'Connor, *The Holy Land: An Oxford Archaeological Guide from Earliest Times to 1700* (Oxford: Oxford University Press, 2008), 330.

it back in order to build there a sumptuous winter palace complex. This was situated along both sides of the Wadi Qelt, roughly two hundred fifty meters to the east of the Hasmonean palace. Most likely built during the last decade of Herod's rule, it encompassed a number of earlier structures. There was an older palace on the south side of the wadi, along with a large pool and an adjacent sunken garden. A chronologically newer segment of the palace lay on the wadi's north bank, being linked by a bridge to the southern complex. It boasted an enormous reception hall, Greco-Roman baths, and two column-lined courts.[13]

Though nothing remains today but foundations and a few fragments of wall decorations, it is clear that these structures, built by enslaved Jewish hands for a Jewish imposter-king, were magnificent beyond words. It is clear that although by now the land of Israel was nothing more than a Roman province, a land bridge of sorts strategically connecting Egypt to the rest of the empire, its population, culture and religion were still authentically Jewish. To intimate that this ancient city of palm trees, the City of the Moon, was devoid of Jewish presence from hoary antiquity, seems preposterous on its face; but such are the realities of today's propaganda-charged climate.

It is fair to ask, in any case, what became of the Jews of Jericho? It seems that after the second great revolt against Rome, the Bar-Kokhba Revolt of 132-135 C.E., the land lay in such devastation that there was little to sustain them. The majority of the population slowly drifted away, predominately to Babylonia and points east, where they established illustrious academies of Talmud and Torah study. For centuries the Byzantines ruled the land, though significant numbers of Jews still remained in places like Jericho. In addition to the Shalom Al Yisrael Synagogue, the Na'aran Synagogue, dating from the Byzantine era and containing its own large mosaic floor, was discovered in 1918.[14] In the seventh century, however, the Byzantines were expelled by the Arabs, under whom Jericho became part of the military district of Palestine. During the Umayyad Dynasty a great "desert castle" was constructed, not far from Jericho, which became known as Hisham's Palace. Though destroyed by an earthquake, it remains a tourist attraction to this day.

Did these Muslim Arabs express any sense of guilt or remorse for having conquered lands belonging to the Byzantine Empire, formerly the Roman Empire, which had in turn conquered it from the Jews? Of

---

13 Magness, op. cit., 184-185.

14 See Dan Urman, Paul V. M. Flesher, *Ancient Synagogues: Historical Analysis and Archaeological Discovery* (Lieden: Brill, 1998), 310.

course not; it would never have occurred to them. This is how peoples and nations have behaved from time immemorial, according to what has been called the "conquest ethic." However, when, many centuries later, Israeli Jews essentially reconquered Jericho, along with the rest of Judea and Samaria in a defensive war, it was considered the greatest of illegal acts, worthy of untold international condemnation. It is nonetheless noteworthy that, unlike other conquerors down through history, at no time did the conquering Israelis expel the Arab population of Jericho or any of the other so-called "West Bank" territories. On the contrary, the government of Israel, in allowing Jews to settle in these territories, also encouraged creative ways of bringing economic prosperity to the Arab population, which, during fifteen hundred years of Islamic rule, had known little but squalor and poverty.

Blackjack in Jericho

In true entrepreneurial spirit, the "occupying" Israelis, who had signed the Oslo Accords in 1993, launched a joint venture with the Palestinian Authority and Jordan to build a Las Vegas-style casino on the outskirts of Jericho. It opened in September 1998.[15] Israelis are known for being among the world's most prolific gamblers; however, gambling is forbidden in Israel. Consequently, a Jericho casino, being a mere thirty minute drive from Jerusalem, seemed a suitable location for them to recreate. The location was also convenient enough for Jordanians and other foreign tourists. Called the Oasis, its construction cost came in at ninety-two million U.S. dollars. Its operators wanted their patrons to forget that they were in Palestinian-controlled Jericho, and they consequently built the casino as close as possible to Israeli territory, just a few hundred yards from the border checkpoints.

On 2,800 square meters of gaming floor, it sported thirty-five tables and two hundred twenty slot machines, all of which were destined to increase with attendant popularity. An adjacent hotel, called the Intercontinental, boasted over two hundred rooms. The upshot of the project was the bringing of jobs and considerable economic stimulus to the greater municipality, in spite of the fact that gambling is strictly forbidden by Islam, just as it is by Judaism. Within a single year of opening, the casino averaged some 2,800 visitors, collectively spending an average of one million dollars per day. The new building brought modern glitz, glitter, and prosperity to the ancient city, becoming the largest private

---

15 Deborah Sontag, "Jericho Journal; Arafat's Gamble: A Casino for an Israeli Clientele," *New York Times*, Sep. 15, 1998.

employer in Judea and Samaria. Unsurprisingly, Yasser Arafat, himself an associate of the Oasis, and his economic advisor Muhammad Rashid, seem to have been involved in money laundering through the casino. In any case, on the heels of the casino's inauguration, a Jericho Resort Village was envisioned, including a golf course, convention center, and a cultural activities center. There was even a plan to construct a cable car between ancient Jericho and the traditional location of Jesus' forty-day fast, the Mount of Temptation.[16]

Jericho Casino

Unfortunately, just as the casino was thriving, the al-Aqsa Intifada erupted, in September 2000. During the first days of the uprising, Palestinian gunman ensconced themselves in the casino, firing on Israeli soldiers. The I.D.F. returned fire, blowing a hole in the front of the facility and forcing its closure. The militants retreated, and while the facility was repaired, it was never reopened.[17] The building was transformed to an empty hulk, and the promised economic benefit to Jericho faded away. It has on occasion been pointed out that if the Palestinian leadership would simply abandon its goal of destroying Israel and instead exploit its geographical position, living next door to the most creative people on earth, their own people would become wealthy and happy. If they would

16 See *Journal of Palestine Studies*, 30/1-2 (Institute for Palestine Studies and Kuwait University, 2000): 16.

17 See P. R. Kumaraswamy, *Historical Dictionary of the Arab-Israeli Conflict* (Lanham, MD: Rowman & Littlefield, 2015), 251.

realize that those who live near to successful entrepreneurs might learn from them and initiate their own entrepreneurial ventures, then the region might be fundamentally transformed for the better. This of course was not to be the case.

Doorway to Jericho

Perhaps the most controversial aspect of Israel's very existence relates to a small but solid portion of its citizenry who have decided to create new facts on the ground by choosing to live and settle in lands won during the 1967 Six-Day War. To them these territories are not the occupied "West Bank" of the Jordan River, but the biblical heartland of Judea and Samaria. It is germane to point out that these Jewish settlers have not moved into these territories indiscriminately, but have carefully chosen where they will reside. In many, if not most, instances that choice is determined by the rich archaeological heritage of the area, the facts underground.

Mitzpe Yerikho

In 1977 a settlement was established, located some twenty kilometers east of Jerusalem, and known as Mitzpe Yerikho ("Lookout of Jericho"). True to its name, it is perched on a desert hill overlooking the Jordan Valley and the ancient "City of Palms" beneath. It has been condemned in the international community as an illegal settlement, the government of Israel being accused of confiscating the Palestinian land on which it was established.[18] Yet, the five hundred religious Zionist families who live here have created homes and infrastructure in a place that was utterly barren and uninhabited. Reclaiming the desert is considered provocative and even criminal. Why? Because Jewish Israelis are doing it. In point of fact, the pioneers who first established the community wanted to live on the outskirts of Jericho itself, but the Israeli government persuaded them to move to a more distant locale so as to avoid provoking the Arab population. As a religious settlement, it is home to Ashkenazi, Sephardi, Yemenite and Chabad synagogues as well as a *yeshivah*, but it has also brought commerce to the area, offering bike trips and jeep tours for visitors anxious to explore the nearby Wadi Qelt and monastery.[19] This of course would not suffice in a land designated by both the Palestinian Authority and the world community as *Judenrein*.

In 1999 another settlement, initially designed as a station for agricultural experiments, a so-called acclimatization farm, was established to the immediate north of Jericho.[20] Its purpose was the cultivation, in this harsh desert environment, of grapes, figs, dates, lemons, passion fruit and sweet potatoes. Having effectively created a Jewish foothold in the region, many families began moving in, naming the new farming community Mevo'ot Yericho. Temporary trailers were replaced by permanent homes. Within a decade the settlers established a special humanitarian project known as Ginat Eden (the female form of "The Garden of Eden"), for disadvantaged girls, who live and work in the settlement by day and travel in the evenings to study in Jerusalem.[21] In 2007, the world's first solar-powered ritual immersion bath (*mikveh*) was put into operation at the settlement. While the government of Israel never officially sanctioned the creation of Mevo'ot Yericho, neither did it object to what was in effect an outpost designed to interrupt the establishment of a contiguous Palestinian state. As expected, the majority of the international community considered it to be one more Israeli provocation, and another impediment to the so-called "peace process."

---

18 See Luke Baker, "Israel's Settlement Drive Is Becoming Irreversible, Diplomats Fear," Reuters, May 31, 2016: https://www.reuters.com/article/us-israel-palestinians-settlements/israels-settlement-drive-is-becoming-irreversible-diplomats-fear-idUSKCN0YM1MY.

19 https://sites.google.com/site/mitzpeyericho/about-mitzpe.

20 "Israel Approves New Legalization of West Bank Settlement Days before Polls," *i24 News*: https://www.i24news.tv/en/news/israel/politics/1568546690-israeli-cabinet-approves-legalization-of-west-bank-settlement-mevo-ot-yericho.

21 http://ginateden.com.

In December 2016, the Obama administration enabled the passing of UN Security Council Resolution 2334, declaring Israeli settlement activity a "flagrant violation" of international law, having "no legal validity." In September 2019, tentative approval was granted by the cabinet of the then-transitional Israeli government to make Mevo'ot Yericho officially legal under Israeli law.[22] The creation of a permanent government had proved to be elusive after the second round of elections that year, and Israel's Attorney General nullified the declaration. Two days prior to the September 17 election, Benjamin Netanyahu promised that the move would be completed in the next government, assuming that his party remained in power. The election results, however, proved just as inconclusive as those of the previous election, and the legalization will remain in legal limbo.

Netanyahu nonetheless announced in September 2019 that if he were to remain Israel's prime minister, he would make use of a "historic opportunity" to extend Israeli sovereignty over the entire Jordan Valley. He declared:

> As much as it is possible, I want to apply sovereignty in the communities and other areas with maximum coordination with the U.S.... But there is one place where it is possible to apply Israeli sovereignty immediately after the election, if I receive a clear mandate to do so from you, the citizens of Israel. In recent months I have led a diplomatic effort in this direction, and the conditions for this have ripened. Today I am announcing my intention to apply, with the formation of the next government, Israeli sovereignty on the Jordan Valley and northern Dead Sea.[23]

The Palestinian side of course condemned the idea, as former head negotiator Saeb Erekat called it "manifestly illegal," adding that it "merely adds to Israel's long history of violations of international law."[24] Condemnation notwithstanding, U.S. Secretary of State Mike Pompeo announced, in November 2019, a stark shift in American policy toward the Israeli settlements in Judea and Samaria. Whereas virtually the entire world community considers all such settlements t o be strictly illegal, the American administration would no longer view them as such.

22 i24 News, op. cit.

23 Noa Landau and Yotam Berger, "Netanyahu Says Israel Will Annex Jordan Valley if Reelected," *Haaretz*, Sep. 10, 2019.

24 Ibid.

Pompeo announced:

> After carefully studying all sides of the legal debate, this administration agrees with President Reagan. The establishment of Israeli civilian settlements in the West Bank is not per se inconsistent with international law.

Prime Minister Netanyahu, commenting on this declaration, said:

> The United States adopted an important policy that rights a historical wrong when the Trump administration clearly rejected the false claim that Israeli settlements in Judea and Samaria are inherently illegal under international law. This policy reflects an historical truth – that the Jewish people are not foreign colonialists in Judea and Samaria. In fact, we are called Jews because we are the people of Judea.[25]

Netanyahu was of course appealing to history, but history bolstered by the "facts underground," which portray a long and uninterrupted Jewish presence in this "contested" region, over the course of millennia. Moreover, they completely contradict the false narrative of a Palestinian Arab presence in "Canaan" for some ten thousand years, dating back to ancient Jericho. Like a proverbial broken record, that claim was reiterated at a UNESCO session on November 15, 2019, when Riyad al-Malki, Palestinian Minister of Foreign Affairs and Expatriates, stated:

> Palestine is the cradle of culture and religions, and part of our Palestinian people's past, heritage and history is engraved on the walls of its capital of Jerusalem – this heritage that Israel, the occupying power, is working to destroy and control, and [it is working] to falsify the history that is witness to our people's rootedness in its land for more than ten thousand years, which refutes the Israeli occupation's settlement colonialism narrative.[26]

By contrast, *Palestinian Media Watch* noted:

> The PA habitually refutes the authenticity of the numerous archaeo-

---

25 Yaakov Katz, "West Bank settlements not illegal, Pompeo announces in historic shift," *Jerusalem Post*, Nov. 18, 2019: https://www.jpost.com/Israel-News/West-Bank-settlements-not-illegal-US-decides-in-historic-US-policy-shift-608222.

26 Donna Rachel Edmunds, "Palestinian Academics Deny Archaeological Evidence of Jews in Israel," *Jerusalem Post*, Nov. 30, 2019: https://www.jpost.com/israel-news/palestinian-academics-deny-archaeological-evidence-of-jews-in-israel-609507.

logical artifacts and non-biblical sources that testify to the Jewish presence and nationhood thousands of years ago.[27]

Lies, like old soldiers, never die; but perhaps, as archaeology serves
to concretize the historical record, they may one day "just fade away."

---

27 Ibid.

CHAPTER SEVEN

REWRITING HISTORY –
SHOWDOWN AT SHECHEM

R AIN WAS PRECIOUS, even sacred, in the ancient Near East, and when, in the winter months, the heavens let loose thunderous torrents, it was time for rejoicing. In places where the rocky, immovable wadis flooded, the thankful people of the land began to cultivate their pastures and tend to their thirsty flocks. Where the lay of the land was suited to travel, sojourners made their way on ground which soon emerged as a thoroughfare. Where spring water emerged from the earth, erstwhile travelers paused to drive their tent stakes, and thus established the rudimentary foundations of a city.

At the northern end of a high road called "The Way of the Patriarchs," traversing the hill country of Ephraim and stretching from Jerusalem to Israel's northern districts, sat such a city, four thousand years old, whose legacy is enshrined in the biblical record. Its name, Shechem, is difficult for most Westerners, and lacks the mystique of places like Jerusalem and Bethlehem; but its prominence in the Bible cannot be overstated. The derivation of the word "Shechem" is anatomical, referring to the "shoulder" or "back." It is well-suited to its locale, lying in a narrow valley between the lofty high places of Israelite lore, Mount Gerizim and Mount Ebal. Today, it is also the location of the Palestinian town of Nablus, a city of about 120,000, and one of the many flash-points in the Arab-Israeli conflict. Mark Twain, on his well-documented, nineteenth-century journey to the Holy Land, remarked:

The narrow canyon in which Nablus, or Shechem, is situated, is under high cultivation, and the soil is exceedingly black and fertile. It is well watered, and its affluent vegetation gains effect by contrast with the barren hills that tower on either side.[1]

View from Mt. Ebal, near Nablus

He further recorded his precise itinerary in his meticulous journal:

At two o'clock we stopped to lunch and rest at ancient Shechem, between the historic Mounts of Gerizim and Ebal, where in the old times the books of the law, the curses and the blessings, were read from the heights to the Jewish multitudes below.[2]

Regarding the Jewish community of Shechem, he wrote:

For thousands of years this clan have dwelt in Shechem under strict taboo and having little commerce or fellowship with their fellowmen of any religion or nationality. For generations they have not numbered more than one or two hundred, but they still adhere to their ancient faith and maintain their ancient rites and ceremonies. Talk of family

---

1   Mark Twain, *The Innocents Abroad*, 322.
2   Ibid.

and old descent! … This handful of old first families of Shechem… can name their fathers straight back without a flaw for thousands [of years]…. I found myself gazing at any straggling scion of this strange race with a riveted fascination, just as one would stare at a living mastodon or a megatherium…[3]

The strategic situation of Shechem was ideal, to the north of the biblical cities of Bethel and Shiloh. It was at the intersection of important trade routes, one between north and south, the other east and west, being essentially able to control traffic in all directions. Its major deficiency, however, was in natural defenses, and for this reason it required major fortifications. Its water supply came from so-called Jacob's Well, a scant four hundred meters to the southeast, and via a conduit from a cave at Mount Gerizim. The verdant fields of the 'Askar plain provided the inhabitants with ample sustenance.

Shechem and Mt. Gerizim from Mt. Ebal

Abraham, Ibrahim and the Politics of Denial
Archaeology reveals that Shechem was a dominant center in this region some three and a half millennia ago. The city is attested by an Egyptian monument, an upright stone slab or "stela," bearing a hieroglyphic inscription of a nobleman in the court of Pharaoh Sesostris III, who ruled from 1878 to 1839 B.C.E. The hieroglyphs read:

---

3 Ibid.

His majesty reached a foreign country of which the name was skmm [Shechem]. Then skmm fell, together with the wretched Retunu [an Egyptian name for the inhabitants of Syro-Palestine].[4]

Nonetheless, the city strongly resisted Egyptian suzerainty. Also uncovered was a clay tablet, bearing curses and ritually broken, known as an execration text, dating to the mid-1800s B.C.E. It references a certain Ibish-hadad of Shechem, who was leading the resistance against Egyptian rule in those days.[5] The name of Shechem's king, Labaya, is also found in cuneiform tablets, dated to the 1300s B.C.E. and discovered in the royal archive at Tel el-Amarna, Egypt. Egyptian troops had been sent to crush a rebellion launched by King Labaya, but failed in their attempt to bring him in line with Pharaonic authority.[6]

Though the name Shechem is largely unsung and uncelebrated, even by avid readers of the sacred texts, its ancient stones echo the tales of the great biblical patriarch, Abraham, whose travels from his birthplace in Babylon were directed by divine command: "Go ye ... to the land which I will show you" (Genesis 12: 1). Shechem, among the oldest cities in Canaan, was one of the places where Abraham camped.

Listed by UNESCO in its Inventory of Cultural and Natural Heritage Sites, Shechem is one more element of "proof" (UNESCO politics notwithstanding) of Jewish claims to the Holy Land. At least the Bible would lead us to think so. We read that Abraham was passing near the city when the land of Canaan was promised to his descendants:

Abram passed through the land to the place of Shechem, to the terebinth of Moreh. And the Canaanite was then in the land. And the LORD appeared to Abram, and said: "To your seed I will give this land"; and he built there an altar to the LORD, who appeared to him (Genesis 12:6-7).

There could be not clearer declaration of the Jewish people's eternal attachment to the biblical heartland. Today's Palestinian Arab narrative is, however, more than a trifle different. While Islam accepts Abraham/Ibrahim as a great prophet, he is not considered the father of the Jewish

---

4 Walter Harrelson, "Shechem in Extra-Biblical References," *The Biblical Archaeologist,* 20/1 (Feb. 1957): 2-10.

5 Lawrence A. Toombs, in David N. Freedman, ed., *Anchor Bible Dictionary,* Vol. 5. (London: Doubleday, 1992), 1179.

6 See George Kufeldt, "Labaya of Shechem and the Politics of the Amarna Age" (1974). *Dropsie College Theses.* 99.

people, but of the Muslims. There never was a Jewish presence in the land, so they say. A Palestinian political science professor at Al-Azhar University, Riyad al-Aileh, opined on regional television program "The Supreme Authority:"

> The Jews claim that they were in Palestine two thousand years ago. If we look at the history, we will see that they were not in Palestine in the past, but rather only as invaders less than seventy years ago. For these seventy years they have been invaders, like the Hyksos, the Byzantines, the Persians and [British] colonialism. The Canaanite Palestinian people have since succeeded in defeating those invaders and continue [to live] in this land.[7]

In another Palestinian television interview, for a program titled "The Scent of History," a member of the Jerusalem branch of Fatah (Abir Zayyad), declared:

> We have no archaeological evidence of the presence of the children of Israel in Palestine in this historical period, three thousand years ago, neither in Jerusalem, nor in all of Palestine.[8]

Novelist Haidar Massad, on the television show "Palestine This Morning," commented:

> I wrote a novel called *The Palace* that was published in 2019. This novel… is about the falsification of the historical geography in the Zionist and Talmudic (i.e. Jewish) narrative… The reader can establish… that in this land, Palestine, which has always been Arab – the children of Israel were never there.[9]

American academic, Zev Garber, observes that writing Jews out of the historical record is pandemic, noting the tourism website of the Palestinian Authority, which claims that Palestine has contributed greatly to human civilization, while completely ignoring the Jewish presence:

> The crucible of prehistoric cultures, it is where settled society, the alphabet, religion, and literature developed, and would become a meeting place for diverse cultures and ideas that shaped the world we know

---

7 Edmunds, op. cit.
8 Ibid.
9 Ibid.

today.[10]

However, as Garber notes, the very name "Palestine" was a moniker coined by the Romans, and the very idea that today's "Palestinians" are the "original" inhabitants of this ancient land is utter nonsense. He queries, "How authentic is the Greco-Roman designation 'Palestine' to a land mass claimed by a so-called Palestinian people two millennia ago, let alone in prehistoric times?"[11]

If, however, the Palestinian claim on the land is not "prehistoric," how exactly did today's Palestinians acquire claim to it? Is it not the case that their ancestors simply moved in and took it? History tells us that those who identify as "Palestinians" are in fact the remnants of sundry Arab tribes living under the Ottoman Turks, who lost the land to the British in 1917. The Ottomans took it from the Mamluks, who took it from the Ayyubids, who took it from the Crusaders, who took it from the Seljuks, who took it from the Fatimids, who took it from the Abbasids, who took it from the Byzantines and Romans, who took it from the Jews. This is exactly what the archaeological record also reveals, in sites like the ancient city of Shechem in the heart of biblical Israel. Today it is essentially a suburb of Nablus, surrounded by Palestinians, who scoff at the notion that ancient Shechem was ever Israelite.

Old City of Nablus with Mt. Gerizim in background

10 http://www.travelpalestine.ps/print.php?id=3b7y951Y3b7.

11 Zev Garber, "The Pittsburgh Shabbat Massacre: Terms of Depiction and Destruction: Old-New Usage" in *Journal of Ecumenical Studies* (2020).

Canaanite Shechem

It is here, in several important excavations, interrupted by two world wars and numerous rounds of Mideast upheaval, that one of the latest controversies in the politics of archaeology comes into focus. Today's Palestinians are using the dig to advance their own agenda, but the story of the excavations in Shechem begins over a century ago, when its ruins appeared to modern eyes for the first time.

The site, known at the time as Tel Balatah, was identified as Shechem in 1903, when a team of German archaeologists began to dig in earnest.[12] Their conclusion, that they had indeed found the biblical city, has held up well to the present day. A second excavation, conducted by an Austro-German archaeological team, was conducted in 1913-1914. Following the First World War, the work resumed, continuing for a full decade, from 1926-1936. Yet another excavation was undertaken in 1956, under renowned archaeologists G. E. Wright and B. W. Anderson, followed by the 1973 dig under William Dever.[13]

Ancient Shechem

Over the course of decades, layer upon layer of remains revealed themselves to the excavators' eyes, seemingly begging to relate their stories – of conquest, destruction, rebuilding, and inhabitation by an assortment of tribes and ethnicities. From the Middle Bronze Age (before Joshua and the Israelites arrived), there were earthen embankments and a substantial wall built of Cyclopean stones. The foundations of a Middle Bronze gate, with three piers and two chambers, bear witness to the city's approach during the second millennium, B.C.E. Might this have been the very threshold crossed by the patriarch Jacob?

Archaeology cannot tell us who comprised the city's earliest residents. Perhaps they were the Semitic people known as the Hyksos, who migrated from northern Syria and who ruled Egypt between the Middle and New Kingdoms. The biblical text makes it clear that the population was Canaanite, to be supplanted by Israelites. There was certainly never a people known as "Palestinians," to whom the land rightfully belongs, as its "earliest residents." Should we perhaps find some Canaanites to whom to return the city? The Canaanites have of course disappeared, as has the entire lineage of the ancient Romans and Byzantines. There are no Egyptians (only Arab tribes who conquered the land of the pharaohs), no Babylonians, no Assyrians, no Edomites, Hittites or Amalekites; and the list goes on. Of all ancient peoples, only the Jews have survived. Yet, it is precisely the Jews who are said to be squatters on Arab land. Unfortunately, today's Middle East is, more often than not, the theater of the absurd.

There is of course no proverbial "Kilroy was here" evidence proving that the biblical patriarchs passed this way. There are, however, remains that are eerily consistent with the stories of Abraham, Isaac and Jacob, as recounted in the book of Genesis. From the period known as Middle Bronze I, there are mud brick walls resting on stone foundations, along with abundant examples of domestic artifacts typical of that time.[14] By no means has anything been found here that does not match the general background of the biblical narratives.

There was, for example, the story of Dinah, who went out among the women of Shechem, where her father Jacob had purchased land and where her clan had encamped. We read:

> And Shechem the son of Hamor the Hivite, the prince of the land, saw her; and he took her, and lay with her, and humbled her. (Genesis 34:2)

The implication is that Dinah was raped. Shechem's father approached Jacob to arrange marriages between their peoples: "Make marriages with us; give your daughters unto us, and take our daughters for yourselves" (Genesis 32:9). The sons of Jacob agreed, on condition that the Shechemites be circumcised. The narrative relates the loathsome details of the brothers' revenge:

> And it came to pass on the third day, when they were in pain, that two

---

14 Toombs, op. cit.

of the sons of Jacob, Simeon and Levi, Dinah's brothers, took their swords, and came unseen upon the city, and slew all the males.

… And Jacob said to Simeon and Levi: "You have brought trouble to me, by making me odious to the inhabitants of the land, to the Canaanites and the Perizzites; and, being few in number, they will gather themselves against me and attack me; and I shall be destroyed, both I and my household." (Genesis 34:25, 30)

Shechem seizes Dinah

Jacob was subsequently directed by divine providence to move on to Bethel, and then to Hebron.

Shechem is next found in the story of Jacob's son, Joseph, dispatched to inquire of his brothers, who were grazing their flocks near the city: "So [Jacob] sent him out of the valley of Hebron, and he came to Shechem" (Genesis 34:14). Joseph is told that his brothers have moved on to Dothan, some distance to the north. It is there that Joseph is treacherously sold into slavery, as prelude to the classic story of his rise to power in Egypt, as well as the four-hundred-year sojourn of the Hebrews in the land of the pharaohs, followed by their deliverance under Moses. We will not hear of Shechem again until the Israelites invade the land of

Canaan under Joshua.

In the twentieth century excavation of the city, Albright's protege, G. E. Wright, uncovered at Tel Balatah the most massive Canaanite structure in Israel. Measuring twenty-one meters wide, twenty-six feet in length, its stone foundation walls, over five meters thick, once supported a great temple of imposing height, built of timber and mud brick. On its flanks were two huge towers, standing sentinel over its entrance.[15] Referred to in Wright's archaeological report as "Temple 1," the site was reexamined in 2003 by Harvard professor Lawrence Stager, who found a potential link with a compelling tidbit in the book of Joshua:[16]

> So Joshua made a covenant with the people that day, and set them a statute and an ordinance in Shechem. And Joshua wrote these words in the book of the law of God; and he took a great stone, and set it up there under the oak that was by the sanctuary of the LORD. (Joshua 24:25-26)

Certainly, the "sanctuary of the LORD" was originally a pagan Canaanite shrine, most likely converted to use as an Israelite temple at some point after the Israelite conquest. Later, in the book of Judges, the city again features prominently, as an element in the narrative of the power-hungry Israelite judge, Abimelech, who sought for himself the title of king. We read:

> And all the men of Shechem assembled themselves together, and all Beth-millo, and went and made Abimelech king, by the terebinth of the pillar that was in Shechem. (Judges 9:6)

However, when the people of Shechem later rose up against the megalomaniacal monarch, he carried out unforgiving vengeance:

> And Abimelech fought against the city all that day; and he took the city, and slew the people within; and he beat down the city, and sowed it with salt. (Judges 9: 45)

The rebel leaders fled to the temple of El-berith (meaning "God of the covenant"):

---

15 See G. E. Wright, "The Samaritans at Shechem," in *Shechem: Biography of a Biblical City* (New York: McGRaw-Hill, 1965) 170-84.

16 See Lawrence E. Stager, "The Shechem Temple," *BAR* 29/4 (Jul.-Aug. 2003).

And when all the men of the tower of Shechem heard of it, they entered into the stronghold of the house of El-berith. (Judges 9:46)

Abimelech's retribution was ferocious and fiery:

Abimelech took an axe in his hand, and cut down a branch from the trees, and took it, and laid it on his shoulder; and he said to the people who were with him: "What you have seen me do – quickly! – do as I have done." And all the people did likewise – each one cutting down a branch – and followed Abimelech, and piled them against the stronghold, and set the stronghold on fire above those within; so that all the people of the tower of Shechem also died, about a thousand men and women. (Judges 9:48-49)

While the details of the atrocity cannot be confirmed by archaeology, G. E. Wright identified the "tower (Hebrew: *midgal*) of Shechem" with "Temple 1."[17] According to Stager, the Hebrew word for "sanctuary" (*mikdash*) referenced in the Joshua account might also refer to the great temple called a *midgal* ("tower") in the book of Judges and ultimately excavated in the early twentieth century.

As the biblical narrative continues, we are told that when Abimelech besieged another nearby city, Thebez, a woman on another high tower hurled down a millstone on his head, thus ending his deceitful ambitions. Such accounts are not lost on modern Israelis, who embrace the core-Jewishness of the whole land of Israel, and certainly not the Palestinian "rewriting" of history.

From the standpoint of chronology, the book of Judges recounts that Abimelech is succeeded by Jephthah, who is commissioned by the Israelites to challenge their Ammonite foes. Jephthah, referring to Moses bringing the Israelites through this region on their way to Canaan, makes known to the Ammonite king that his people have been living in the territory east of the Jordan River for some three centuries. Since archaeology attests to the destruction of Canaanite Shechem (presumably at the hands of Abimelech) having occurred around 1100 B.C.E., we can, moving backwards, place the date of Joshua's conquest of the land to roughly 1400 B.C.E. This coincides remarkably well with the traditional biblical chronology of the exodus from Egypt at 1446 B.C.E. and the invasion of Canaan at 1406 B.C.E.

---

17 Wright, op. cit.

The Death of Abimelech by Gustave Doré

Israelite Shechem

We know very little about Shechem in the period of the Israelite monarchy, including the reigns of the first kings, Saul, David and Solomon. We are told, however, that after the death of the wise King Solomon, his foolish son Rehoboam takes the throne at Shechem, where all the tribes had assembled, announcing: "Whereas my father burdened you with a heavy yoke, I will add to your yoke; my father chastised you with whips, but I will chastise you with scorpions" (I Kings 12:11). In response, ten of Israel's twelve tribes revolted, pulling away to form their own kingdom in the north (including Shechem) and choosing their own king, the wicked Jeroboam I. This northern kingdom would be called Israel, while the kingdom of the south would be known as Judah. All, however, should be considered Israelites, from whom today's Israeli

Jews proudly claim descent.

The archaeological remains corresponding to the reign of Jeroboam and his successors appear in Levels X and IX at Tel Balatah (the topmost layer being Level I), where the foundations of carefully constructed stone houses have been discerned. How can it be assumed that these remains are Israelite rather than Canaanite? The telling answer is in the foundations themselves, where there is evidence of stairs leading up to a second story, a characteristic of four room houses, which, as noted by William Dever, are often seen as archaeological "markers" of Israelite habitation.[18] Archaeologist Edward F. Campbell specifically pointed to Level IX, dating from 920 B.C.E. to 810 B.C.E., citing 1 Kings 12:25: "Then Jeroboam built Shechem in the hill-country of Ephraim, and dwelt there." He noted: "tangible evidence of Jeroboam I's rebuilding and a return to city status."[19]

In the year 724 B.C.E. the city was laid waste again, this time by the conquering army of the Assyrian empire. In Level VII, it was discerned that Shechem was "reduced to a heap of ruins, completely covered by debris of fallen brickwork, burned beams and tumbled building stones."[20] The Assyrian juggernaut was so thorough that, in addition to the destruction of entire cities, foreigners from surrounding lands were resettled in the surrounding territory, so as to obliterate not only the Israelite kingdom, but the ethnic population as well.

This was the origin of the Samaritan sect, who mingled elements of Judaism with their own traditions. Shechem experienced a comeback of sorts during the Hellenistic age, from the fourth through second centuries B.C.E., during which time many fine buildings were constructed. The Samaritans built their own temple on nearby Mt. Gerizim, declaring that the tables of the Ten Commandments, given by Moses, had been deposited there, along with the "true" version of the Torah. While the temple itself was destroyed in antiquity, the Samaritan sect has survived to this day, performing a yearly animal sacrifice on Mt. Gerizim and faithfully preserving the so-called Samaritan Penteteuch.

---

18 William G. Dever, "Monumental Architecture in Ancient Israel in the Period of the United Monarchy," *Bible and Spade* 7: 80–81.

19 Campbell, op. cit., 1352–53.

20 Toombs, op. cit., 1185.

Samaritans on Mt. Gerizim

Shechem experienced yet another decline when two rival dynasties, the Ptolemies of Egypt and the Seleucids of Syria, battled each other and pushed their territorial claims back-and-forth across the Levant. The Jewish people, however, could not be permanently suppressed, nor their claims on the land the squelched. In the year 168 B.C.E., a full scale revolt against Syrian suzerainty broke out at the hands of a group of Jewish freedom fighters, famously known as the Maccabees. During the course of four years they defeated army after army of Syrian troops, liberating the city of Jerusalem and rededicating the temple which the Seleucids had defiled. To commemorate their victory, they established a great feast, celebrated to this day as Hanukkah. From Jerusalem, they went on to press Jewish sovereignty across the whole of the ancient land, including Shechem. One of the Maccabee clan, John Hyrcanus, razed the Samaritan temple to the ground around 126 B.C.E., and destroyed Shechem yet again in 107 B.C.E. The city never recovered, and its original site lay in silent repose until the first of the archaeological excavations, at the beginning of the twentieth century.

It was in the first century B.C.E. that the Romans turned the land into an outlying province, building their own Greco-Hellenistic cities adjacent to old Israelite settlements. A new city located about one mile (1.6 km.) west of Tel Balatah was constructed in 72 C.E. and named Flavia Neapolis.[21] The word "Neapolis" ("new city") lay at the root of the Arabic name for the town – Nablus – but of course the Arabs did not move in until the Byzantines were forcibly ejected. The Samaritan sect continued to thrive in the region under the Romans, as evidenced by

---

21 Itzhak Magen, "The Sacred Precinct on Mount Gerizim," *Bible and Spade* 14/40 (2001).

human burial remains dating to this period discovered on the low slopes of Mt. Ebal.[22] However, their attempts to restore their cultic worship on Mt. Gerizim were suppressed.

Entrance to Jacob's Well

Jacob's Well

An ancient well located about five hundred yards (460 m.) to the southeast of Tel Balatah and just south of the Arab village of Askar is believed by to have been dug by the patriarch Jacob himself. Though not mentioned in the Hebrew Bible, multiple faith traditions – Jewish, Samaritan, Christian and Muslim – hold to its authenticity.[23] In the Christian gospels Jesus was said to have visited a woman near "Jacob's well," near Sychar (most likely the same as "Askar"), underscoring the veneration of the site for the last two millennia.[24] Jacob was of course an Israelite, and while non-Jewish faiths have attempted to co-opt his persona, the tradition that it was he who dug this well has left an indelible imprint on the psyche of modern Israelis, who see the Jewish claim on

---

22 Itzhak Magen, "Neapolis," in Ephraim Stern, ed., *New Encyclopedia of Archaeological Excavations in the Holy Land* 4 (New York: Simon & Schuster, 1993), 1358–59.

23 Zdravko Stefanovic, "Jacob's Well," in *The Anchor Bible Dictionary* Vol. 3, David N. Freedman ed. (New York: Doubleday, 1992), 608.

24 See John 4:4-6.

this land as irrevocable and eternal. Of this site Mark Twain also wrote:

> In this same "parcel of ground" which Jacob bought of the sons of Hamor for a hundred pieces of silver, is Jacob's celebrated well. It is cut in the solid rock, and is nine feet square and ninety feet deep. The name of this unpretending hole in the ground, which one might pass by and take no notice of, is as familiar as household words to even the children and the peasants of many a far-off country. It is more famous than the Parthenon; it is older than the Pyramids.[25]

Homegrown Critics

It has nonetheless been charged that despite the long history and tradition behind the archaeology of this region, it is the Israelis, not the Palestinians who are intent on using archaeology politically. Surprisingly enough, the accusation is wielded by a homegrown Israeli organization called Emek Shaveh ("Equal Valley"). Its website describes itself as "… an Israeli NGO working to defend cultural heritage rights and to protect ancient sites as public assets that belong to members of all communities, faiths and peoples." It declares:

> We view heritage sites as resources for building bridges and strengthening bonds between peoples and cultures and believe that archaeological sites cannot constitute proof of precedence or ownership by any one nation, ethnic group or religion over a given place.[26]

It nonetheless features articles which are harshly critical of Israel's archaeological practices. One such article argues that it is Israel which is in fact "appropriating the past." It argues that Israel has, since the 1967 Six-Day War, deliberately made use of archaeology in order to strengthen its iron grip on the conquered territories, namely, Judea and Samaria. As the article puts it:

> The archaeological activity is intended to prove and to strengthen the historical, religious and cultural affinity of the Jewish people and the State of Israel to the West Bank in an attempt to appropriate history and efface the heritage and historical narratives of other peoples and cultures.[27]

25 Mark Twain, *The Innocents Abroad*, Chap. 52.
26 https://alt-arch.org/en/about-us/.
27 http://alt-arch.org/en/wp-content/uploads/2017/12/Menachsim-Eng-Web.pdf.

Specifically, it points to Israel Harel, who was chairman and one of the founders of the umbrella organization of Israeli settlers, known as the Yesha Council (short for "Judea and Samaria" in Hebrew). In a 1981 memo to Israel's Education Ministry, subsequently passed on to the office of Prime Minister Menachem Begin, Harel suggested a number of steps at archaeological locations in the "Occupied Palestinian Territories." These would "… ensure that the Jewish people are in control of the sites which embody its history, its memories and the most obvious and direct testament to its roots and right to the land." Emek Shaveh alleges that the Israeli government has endeavored, since the beginning of the "occupation," to assert ownership of archaeological sites in the "West Bank."

We would do well to ask, however, whether Israeli Jews, who seek to re-inhabit their ancient homeland, should be forbidden from pointing to the evidence beneath their feet of its inherent Jewishness, from time immemorial. Moreover, when the Palestinian side claims repeatedly that there has never been a historical Jewish presence in the land called Israel, is it not understandable and even appropriate to insist that there is a proper role for archaeology in the larger debate about whose "Holy Land" this is? It might in fact be more appropriate to ask whether the founders of Emek Shaveh are themselves being political when denouncing what they perceive as the Israeli expropriation of Palestinian property in the so-called "Occupied Territories." Are they perhaps attempting to assuage their own guilt complex for being citizens of the Jewish state, which, after all, won a war designed to drive them into the sea? Perhaps they are attempting to prove themselves to be morally superior by condemning the practices of their own country, practices which any proud people who had survived their attempted annihilation would certainly undertake.

Emek Shaveh asserts that archaeology is a science, and that it cannot and should not be co-opted for political purposes. While the argument appears cogent, even noble, on its surface, it is certainly the case that all science is subject to interpretation, and to insist that its interpreters be "neutral" is and has always been an impossible standard. The question is not whether archaeology "proves" a people's legacy and attachment to a piece of land, but to which people that attachment historically belongs. To suggest that such a question is an improper interpretation of the scientific data emerging from the archaeological sites is to strip away the fundamental value of the data itself. By contrast, it stands to reason that everything in history has consequences – elections, wars, and even the "unconsidered trifles" emerging from the desiccated dust of some long-forgotten archaeological site. Moreover, the consequences for modern Israelis, struggling to assert the right to live in their ancient homeland, directly on top of the ruins of the past, could not be more germane to today's geopolitical climate.

Shoveling in Shechem

The Israeli-born self-criticism of Emek Shaveh should also be understood against the backdrop of the rancid politics of the region, and the Oslo peace accords of 1993, which handed direct control over Judea and Samaria to the Palestinian Authority. Nablus was transformed into what Israelis considered a "den of martyrs." It became the locus of the so-called Second Intifada ("uprising") against Israeli control, dragging on relentlessly from 2000-2005. Once known as the "Lady of Palestine," Nablus dispatched more suicide bombers, fitted with explosive vests, than any other Palestinian city, in a brazen attempt to slaughter as many Jewish civilians as possible.

In response, the Israel Defense Force launched a major incursion into the "West Bank," known as Operation Defensive Shield.[28] In two days Israel extended its operation northward, past Ramallah, to Nablus. In short order, the city was occupied by Israeli tanks and ground troops. There was heavy street fighting, especially in the city's "Casbah," where explosive devices were set in alleyways and where snipers took up positions. Israeli paratroopers were sent in, and on April 8, 2002, the Palestinians accepted terms of surrender.[29] To this day the city has never fully recovered, and visitors are rare.

When it comes to contemporary excavations in Shechem, Palestinian and Dutch archaeologists a number of years ago turned a garbage dump and junkyard, indiscriminately littered with automobile parts, into an active dig and opened it to the public as an archaeological site. There is, indeed, a not-so-hidden agenda in all of this. Since the West Bank city of Nablus is no longer under direct Israeli jurisdiction, the project, officially conducted by the Palestinian Department of Antiquities, seeks to expose the Arab citizens of Nablus, who have for decades been plagued by violence, bloodletting and isolation, to the richness of the archaeological legacy in their midst. Co-director of the project,

---

28 See Avi Issacharoff and Amos Harel, "Recollections of Israel's Operation Defensive Shield, Ten Years Later," *Haaretz*, Mar. 30, 2012: https://www.haaretz.com/1.5209889.

29 See Amos Harel and Avi Isacharoff, *The Seventh War* (Hebrew) (Tel-Aviv: Yedioth Aharonoth Books, 2004), 251-3.

Leiden University's Gerrit van der Kooij, observed:

> The local population has started very well to understand the value of the site, not only the historical value, but also the value for their own identity. The local people have to feel responsible for the archaeological heritage in their neighborhood.[30]

The city wall and gate of Tell Balatah

Is van der Kooij saying that digging up an ancient Israelite site somehow reinforces "Palestinian" identity? Why are the Palestinians, whose Department of Antiquities is a scant two decades old, suddenly interested in ancient Shechem? The Department's director, Hamdan Taha, stated, "All of the periods in local history, *including that of the biblical Israelites*, are part of Palestinian history."[31] Of course he never addresses how "Palestinian history" can predate the Palestinians. At least Jewish archaeologists are careful not to call pre-biblical finds (including Phoenician, Canaanite, Philistine, etc.) Israelite. Once again, we have a Palestinian attempt to hijack the material remains of the people with whom they are at war. While claiming that Jews have "fabricated" their history, they are fabricating their own. Taha went on, saying that such excavations "give Palestinians the opportunity to participate in writing or re-

---

30 "Archaeologists Uncover Ruins of Biblical City Shekem in War-Torn Palestine," *Fox News*: https://www.foxnews.com/science/archaeologists-uncover-ruins-of-biblical-city-shekem-in-war-torn-palestine.

31 Ibid.

writing the history of Palestine..."[32] In other words, propaganda served by archaeology is not only acceptable; it is essential.

At the end of January 2020, the Trump administration unveiled its "Deal of the Century," an audacious plan aimed at bringing about lasting peace in the Middle East. Its provisions included a sovereign state for the Palestinians, located on seventy percent of the West Bank territory, in return for a renunciation of terrorism and a formal recognition of Israel as a Jewish state. Nablus and the archaeological sites nearby are of course in the heart of the territory to be designated as a Palestinian state. In an era when archaeology is subsumed by politics, it is appropriate to ask: if and when an independent Palestinian state is ever born, what if anything of the Israelite heritage of Shechem will be highlighted or even acknowledged? The facts underground at ancient Shechem will always be present, but their voice will be filtered through the Palestinian propaganda machine. Jacob's Well will still give water, but no Jews will drink from it. Emek Shaveh is right about one thing. When archaeology serves propaganda – Arab propaganda – archaeology as a science disappears.

32 Ibid.

CHAPTER EIGHT

DEFACING HISTORY –
THE TOMB OF JOSEPH

N ESTLED IN THE RUGGED hill country of Samaria sits a sacred shrine, consisting of a squat, plastered dome covering an un-impressive rectangular limestone structure, fronted by a large arched entryway. It is celebrated by three religions (Judaism, Christianity and Islam) as the authentic tomb of the biblical patriarch Joseph. To be sure, no name in world literature evokes more powerful a sentiment than that of Joseph, treacherously sold into slavery by his own brothers, only to rise to prominence as chief courtier to the pharaoh of Egypt. Today Jo-seph's Tomb is situated in the heart of the modern city of Nablus. A century ago it lay in lonely isolation, on a stone-strewn road skirting the foot of a barren mountain. As with most of the "land of milk and honey" celebrated in biblical lore, the place was forlornly barren, desiccated and denuded of trees, with only sparse shrubs clinging to the mountainside.

Mark Twain, on his classic journey to the Holy Land in 1867, duti-fully recorded his impressions of the site:

> About a mile and a half from Shechem we halted at the base of Mount Ebal before a little square area, inclosed by a high stone wall, neatly whitewashed. Across one end of this inclosure is a tomb…. It is the tomb of Joseph. No truth is better authenticated than this. When Jo-seph was dying he prophesied that exodus of the Israelites from Egypt which occurred four hundred years afterwards. At the same time he exacted of his people an oath that when they journeyed to the land of Canaan they would bear his bones with them and bury them in the ancient inheritance of his fathers. The oath was kept. "And the bones

of Joseph, which the children of Israel brought up out of Egypt, buried they in Shechem, in a parcel of ground which Jacob bought of the sons of Hamor the father of Shechem for a hundred pieces of silver." Few tombs on earth command the veneration of so many races and men of diverse creeds as this of Joseph. "Samaritan and Jew, Moslem and Christian alike, revere it, and honor it with their visits. The tomb of Joseph, the dutiful son, the affectionate, forgiving brother, the virtuous man, the wise Prince and ruler. Egypt felt his influence — the world knows his history."[1]

Joseph's Tomb, 1903

Digging Up Joseph

Such an indelible history it was. The story of Joseph is one of the most poignant ever recorded in world literature. Among all the patriarch Jacob's twelve sons, he is the one favored by his father, who bestows on him a splendid, woolen garment, oft described as a striped, multi-colored coat. His eleven siblings, consumed with jealousy, seize him one day, throw him into a pit, and subsequently sell him into slavery, to a camel caravan. They return his coat, smeared with goat's blood, to Jacob, declaring that he has been mauled to death by an animal. The patriarch goes into deep mourning, while Joseph is taken down to Egypt and sold to a wealthy man named Potiphar, one of the pharaoh's officials.

Joseph rises to prominence in his new household, only to become the object of attention of Potiphar's wife, who attempts in vain to seduce

---

1 Mark Twain, *The Innocents Abroad*, Chap. 52.

him. Feeling scorned, she tells her husband that Joseph has attempted to rape her, whereupon the hapless Hebrew is sent away to prison. While languishing in bitter confinement, he successfully interprets the dreams of his fellow prisoners, gaining the reputation of being among the greatest of soothsayers. His fortunes change again, when the pharaoh himself is troubled by inexplicable dreams. Seven healthy, fat cows are eaten by seven thin, emaciated cows; yet, they grow no fatter. Seven lush ears of grain are consumed by seven lean ears, but they remain lean.

When the greatest of Egypt's wise men are unable to decipher the dream, Joseph is summoned from prison to offer his own interpretation. Egypt is about to experience seven years of plenty, followed by seven years of horrible famine. Grain needs to be stored up immediately, in preparation for what is to come. The amazed pharaoh puts Joseph in charge of filling the silos, making him *vizier* of all Egypt. The famine arrives, just as Joseph had predicted, and who should arrive in Egypt in search of food but his eleven deceitful brothers, having come down from Canaan? Joseph now turns the tables on his siblings, who do not recognize him, by planting a silver cup in the grain sack of the youngest, Benjamin, only to have them stopped on their return journey and their grain sacks inspected. Benjamin is accused of theft and threatened with slavery, whereupon Joseph's brother Judah pleads and offers to be made a slave in Benjamin's stead.

Only then does Joseph reveal his true identity, culminating in an unrivaled scene of contrition and reconciliation. He invites his brothers and their families, along with his elderly father Jacob, to come down and dwell in Egypt, in the land called Goshen. At the end of Joseph's life, he insists that when he dies his bones should be brought back to his home in the land of Canaan. According to Jewish tradition, not only Joseph, but his two sons, Ephraim and Manasseh, were interred at the tomb near Shechem.

Does archaeology substantiate any of this celebrated tale? We start with the excavation of the archaeological site in the Nile Delta, the biblical land of Goshen, known as Tell el-Daba, the ancient Egyptian name of which was Rowaty. The site betrays signs of the presence of Asiatics, i.e., Semites, living there as early as the nineteenth century (or 1800s) B.C.E. That happens to be the same time frame in which Joseph was said to have come down to Egypt, along with his family.

Excavations at Rowaty have revealed the remains of rectangular huts made of sand bricks, in which the residents lived. A house of distinctly Semitic style, characteristic of Northern Syria, was also discovered at the site, dating to Egypt's Twelfth Dynasty, specifically the reign of Sesostris III. It is the sort of house in which the biblical patriarch Jacob might have lived, after he came down to Egypt at the behest of Joseph. Later, during the Thirteenth dynasty, there are archaeological remains of an Egyptian palace, apparently the home of an official of some kind. However, its owner seems not to have been Egyptian at all. Interestingly enough, the portico of the structure was supported by twelve pillars. Might they somehow be a reference to Jacob's twelve sons?

Tell el-Daba

It has in fact been suggested that the villa and the surrounding semi-circle of two-room houses are nothing less than the homes of Joseph and his brothers. In the same immediate area we find an ancient cemetery, with twelve graves and artifacts linking the tombs to the houses. One of the tombs was quite substantial, and of pyramidal shape, notwithstanding that only pharaohs and queens were ever buried in pyramid tombs in Egypt. Within it were found the remains of a large statue, with a mushroom-shaped hairstyle, identifying him as an Asiatic, and over his right shoulder a throw stick – the Egyptian hieroglyph for a foreigner. Interestingly enough, only a few bone fragments remained, and no intact skeletons, as with other tombs in the area.[2] A coincidence perhaps, but the biblical text does record Joseph's command that his bones

---

2 See Bryant G. Wood, "The Sons of Jacob: New Evidence for the Presence of the Israelites in Egypt," *Associates for Biblical Research*, Jan. 28, 2016: https://biblearchaeology. org/research/patriarchal-era/3317-the-sons-of-jacob-new-evidence-for-the-presence-of-the-israelites-in-egypt.

be brought out of Egypt and into the Land of Promise. We read:

> And Joseph said unto his brethren: "I am dying; but God will surely remember you, and bring you up out of this land to the land which He swore to Abraham, to Isaac, and to Jacob... God will surely remember you, and you will carry up my bones from here." (Genesis 50:24-25).

The children of Israel remembered their oath, and when they left Egypt during the Exodus, Moses took Joseph's bones with him. This is how they came to be interred near Shechem, at today's West Bank city of Nablus.

Missing Person, Missing Bones

While the Bible is presented as religious history, Mark Twain was of course a writer of fiction, and the question that arises is whether the tomb he confidently declared to be that of Joseph is anything more than a construct of the pious imagination of people long ago. Were the bones of Joseph indeed laid to rest in this hallowed ground? In the halls of academe skepticism reigns, and minimizing the biblical record is ever in vogue. It is argued that solid archaeological evidence connecting the tomb with the biblical Joseph is entirely lacking. It is even argued that the tomb was an ancient Samaritan site, due to the lack of historical sources pre-dating the fourth century of the Common Era. Minimalists suggest that the sublime story of Joseph and his brothers, that builds upon universal themes, of betrayal, forgiveness and redemption, must have been a literary invention, perhaps a novella rooted in Egyptian pagan mythology and cleverly appropriated by Israelite scribes, a millennia and a half later, during the Persian period. Joseph himself becomes a "missing person," at least as far as history is concerned.

What of the biblical tradition that long after Joseph's death, his bones were transported from Egypt to the land of Canaan? The book of Joshua recounts:

> And the bones of Joseph, which the children of Israel brought up out of Egypt, they buried in Shechem, in the parcel of ground which Jacob bought from the sons of Hamor the father of Shechem for a hundred pieces of silver; and they became the inheritance of the children of Joseph. (Joshua 24:32)

Obviously, if there had been no Joseph, there would have been no burial, and, proverbially, "no bones about it." It has even been asserted that the tomb itself could not be more than a few hundred years old. Not surprisingly, there are the "deniers." True to form, there have been arguments from Palestinian sources that the much-hailed tomb has no connection with Joseph, and is not even Jewish. The Palestinian claim is that the biblical Joseph was not a Jew at all, but a Muslim. It is also alleged that the tomb houses the remains of an important Muslim cleric by the name of Yussef al-Duwaik, who was buried there a mere two and half centuries ago. Oddly, the claim has been echoed by some Israelis, who oppose their own state's "occupation" of the "Palestinian territories."[3]

There is in any case no evidence that Muslims ever considered this an important holy site. Moreover, propaganda of this sort flies in the face of documentary evidence, indicating that Jews as well as Samaritans (an offshoot of Judaism) indeed worshiped at this location for more than a thousand years. One of Israel's leading archaeologists, the late Dr. Zvi Ilan, vouched for the site, calling it: "... one of the tombs whose location is known with the utmost degree of certainty ... based on continuous documentation since biblical times."[4]

Joseph Sitings

For example, one early Christian source known as the "Jerusalem Itinerary," or Itinerarium Burdigalense, chronicles the journey to the Holy Land of a certain "Pilgrim of Bordeaux" in the years 333 and 334 of the Common Era. It declares:

At the foot of the mountain itself, is a place called Sichem. Here is a tomb in which Joseph is laid, in the parcel of ground which Jacob his father gave to him.

An early rabbinic commentary on the book of Genesis, dating from 400 to 450 C.E. and known as *Genesis Rabba*, declares:

There are three places regarding which the nations of the world cannot taunt the nation of Israel and say, "you have stolen them." They are: the Cave of the Patriarchs [in Hebron], the Temple Mount and the burial site of Joseph.[5]

---

3 See Daniel Dor, *Intifada Hits the Headlines: How the Israeli Press Misreported the Outbreak of the Second Palestinian Uprising* (Bloomington & Indianapolis: University of Indiana Press, 2004), 48.

4 Zvi Ilan, *Tombs of the Righteous in the Land of Israel* (Israel: Carta, The Israel Map & Publishing Company Limited, 1997), 365.

5 *Genesis Rabbah* 79:7.

The passage goes on to note that each one of these locations was bought "for its full price," by Abraham, by David, and by Jacob. The haunting words of this text could not be more germane to today's Middle East, since the Jewish people are indeed accused of "stealing" these sacred locations.

The fourth-century Christian historian, Eusebius, subsequently commented:

> Suchem, City of Jacob now deserted. The place is pointed out in the suburbs of Neapolis. There is a rumor of Joseph is pointed out nearby.[6]

The Christian Saint Jerome, who lived in the fourth and fifth centuries, personally resided in Palestine for many years and insisted that all twelve Israelite patriarchs were buried near "Sychem." He also cited the pilgrimage of the Roman saint and Desert Mother, Paula, writing: "Turning off the way, she saw the tombs of the twelve patriarchs."[7] That was clearly an exaggeration, since tradition has it that Abraham and Sarah, Isaac and Rebecca, and Jacob and Leah were buried in Hebron. It nonetheless bears witness to the tomb being there and to Joseph, as one of the patriarchs, being in it.

Jerome also recorded that under two Byzantine emperors, Theodosius I and Theodosius II, an expedition was dispatched to remove the bones of the patriarchs to the Hagia Sophia in Byzantium.[8] On failing to recover the remains from Hebron, they traveled north to Shechem, where they forced the local Samaritans to dig up an ancient sepulcher, which they purposed to transport back to Byzantium's great cathedral. However, when they attempted to recover what were deemed to be Joseph's actual bones, a pillar of fire shot up from the ground, forcing the erstwhile grave robbers to flee in terror. Thereafter, the savvy Samaritans piled earth over the site to prevent its further desecration.

Moving into the sixth century, there is an account from another Christian pilgrim and church cleric, a certain Theodosius, stating:

---

6 C. Umhau Wolf, ed., *The Onomasticon of Eusebius Pamphili: Compared with the Version of Jerome and Annotated* (1971), 76-252: http://www.tertullian.org/fathers/eusebius_onomasticon_03_notes.htm#807.

7 Victor Guerin, *Description Geographique, Historique Et Archeologique de La Palestine*, Vol. 1. (Paris: Imprimerie Nationale, 1874), 889.

8 Hagith Sivan, *Palestine in Late Antiquity* (Oxford: Oxford University Press, 2008), 115, n. 24.

"Close to Jacob's Well are the remains of Joseph the Holy."[9] Also dating to the sixth century is the famous Madaba Mosaic, part of the floor of the early Byzantine church of Saint George, located in Transjordan and containing the most important mapped depiction of the land of Israel and the ancient city of Jerusalem that has ever come to light. One of the Greek entries on the map indicates "Joseph's," and coincides with the belief that the patriarchs' tomb could indeed be found in this area around Shechem.

From the twelfth century, there is a curious note recorded in the journal of a traveler from Persia named al-Harawi:

> There is also near Nâblus the spring of Al Khudr (Elijah), and the field of Yusuf as Sadik (Joseph); further, Joseph is buried at the foot of the tree at this place.[10]

Additionally, famed twelfth-century Jewish traveler, Benjamin of Tudela, declared that the Samaritans of Nablus possessed Joseph's tomb.[11] Leading English historian of the twelfth century, William of Malmesbury, described it as being covered in white marble and situated adjacent to the tombs of his brothers.

In the early thirteenth century, Islamic geographer Yaqut al-Hamawi noted:

> There is here a spring called 'Ain al Khudr. Yûsuf (Joseph) as Sadik – peace be on him! – was buried here, and his tomb is well known, lying under the tree.[12]

The French Jew, Menachem Ben Peretz, who lived several years in Hebron during the thirteenth century, wrote that he personally witnessed the presence in Shechem of the tomb of Joseph, son of Jacob, with a marble column at its head and another at its foot. He described it as being enclosed by a low wall. Fourteenth-century Jewish physician, geographer and traveler, Ishtori Haparchi, referenced it at a distance of

---

9 Jonathan Golden, "Targeting Heritage: the Abuse of Symbolic Sites in Modern Conflicts," in Yorke M. Rowan and Uzi Baram, eds., *Marketing Heritage: Archaeology and the Consumption of the Past* (Lanham, MD: Rowman Altamira, 2004), 183–202.

10 See Guy Le Strange, *Palestine Under the Muslims: A Description of Syria and the Holy Land from A.D. 650 to 1500* (New York: Cosimo, Inc., 2010 [1890]), 416.

11 See Adolf Asher, ed., *The itinerary of Rabbi Benjamin of Tudela*, Vol. 1 (London and Berlin: A. Asher & Co., 1840), 67.

12 Ibid., 512.

four hundred fifty meters to the north of Balatah.[13]

Multiple accounts of the tomb's existence are preserved during the centuries which followed, and in spite of various inconsistencies regarding its precise location, the tradition that Joseph's venerated resting place still existed is well attested. A Hebrew inscription on the tomb itself memorializes the repairs made in the year 1749:

> With the good sign. The LORD endures forever. My help comes from the LORD who made heaven and earth. Joseph is a fruitful bough. Behold a renewed majestic building ... Blessed be the LORD who has put it into the heart of Elijah, the son of Meir, our rabbi, to build again the house of Joseph in the [Hebrew] month Sivan, in the [Jewish] year 5509.[14]

בסט ינו עמיעשו ״ בן פורת יוסף ״ לכו חזו בנין מפואר
מחורש חליזו ברוכ ח אשר נתן בלב אליהו בן מאיר רבינו חי י
לבנות שנת את בית יוסף בחורש סיון שחתקט ״ חכותב מאיר
בין יומף מזרחי סט.

Hebrew Inscription at Joseph's Tomb

Nineteenth-century Irish writer and journalist, William Cooke Taylor, observed:

> The present monument is built in the ordinary style of an easter Welee, and is a place of resort, not only for Jews and Christians, but Mohammedans and Samaritans; all of whom concur in the belief that it stands on the veritable spot where the patriarch was buried.[15]

A nineteenth-century Jewish traveler noted the description of Joseph's burial plot in biblical Hebrew as a "portion of field," suggesting a flat area. He observed: "In the whole of Palestine there is not such another plot to be found, a dead level, without the least hollow or swelling in a circuit of two hours."[16]

In other accounts, it was observed that the tomb was often visited by Jews who left multiple Hebrew inscriptions on the walls. The tomb was said to be "... kept very neat and in good repair by the bounty of Jews who visited it."[17] According to another account, the site is described as follows:

> ... a small solid erection in the form of a wagon roof, over what is sup-

---

17 Bishop Alexander, "The Well of Jacob and the Tomb of Joseph," *The Church of England Magazine* 17 (October 26, 1844): 280.

posed to be the patriarchs grave, with a small pillar or alter at each of its extremities, sometimes called the tombs of Ephraim and Manasseh, and the middle of an enclosure without a covering. Many visitors names, in the Hebrew and Samaritan characters, are written on the walls of this enclosure.[18]

In 1878 British soldier and explorer, Claude Conder, described the tomb:

A stone bench is built into the east wall, on which three Jews were seated at the time of our second visit, book in hand, swinging backwards and forwards as they crooned out a nasal chant—a prayer no doubt appropriate to the place.[19]

Joseph's Tomb, interior

Such evidence of the tomb's authenticity is of course anecdotal. There exists no "proof" that would hold up in a courtroom. Scholars would be wise, however, to refrain from dismissing sacred traditions out of hand when they have little better with which to replace them but guesswork. They need only be reminded of British archaeologist Kenneth Kitchen's oft-quoted adage, "Absence of evidence is not evidence of absence."[20] For their part, the ancient Israelites constructed no great pyr-

---

18   Wilson, 60-61.

19   Claude Reignier Conder, *Tent Work in Palestine* (Frankfurt am Main: Outlook, 2018), 44.

20   Kenneth Kitchen, "The Patriarchal Age: Myth of History?" *BAR* (March/April 1995): 50.

amids, no royal mausoleums, no gold-encrusted sarcophagi in which to lay their honored dead. Perhaps the most telling contribution to human civilization was that they fixated themselves on life, not death. To this day, the Jewish prayer for the dead, the Mourners' Kaddish, does not mention death at all, but extols the greatness of God. Their leaders were of a different order than the great pharaonic overlords of Egypt, being defined as much by human frailty as by feigned omnipotence. In life they were authentic, in death humble. It should therefore surprise no one that their burial sites are not as easily identifiable as those of Ramses the Great.

What is clear, however, is that the Jewish people have revered the resting places of their patriarchs throughout the centuries and without break, and that their attachment to specific holy sites long predates the claim of any other people on earth today. Joseph's Tomb, notwithstanding debates about its historical authenticity, is considered the third most sacred site in Judaism, after the Temple Mount and the Tomb of the Patriarchs in Hebron. Geopolitics, however, was destined to sever, at least temporarily, its historical and uninterrupted connection with the Jewish people.

Joseph Jitters

When Israel's war of independence erupted in 1948, the new Jewish state miraculously survived, but Judea and Samaria, the Jordan River's West Bank, fell to the Arabs. The territory was formally annexed by the Hashemite kingdom of Jordan, and for the next two decades, Jews had no access to Joseph's Tomb, Hebron, or the Temple Mount. With Israel's sudden victory in June 1967, however, Joseph's Tomb was back in Jewish hands. Jews began returning to the site on a regular basis, even though Israel refrained from annexing any of the conquered territories. Jews were therefore not allowed to settle in the area. Over time, several Jewish settlements cropped up on the mountainous terrain in the vicinity, including Elon Moreh, Har Brachah, Kedumim and Yitzhar.[21] In the mid-1980s a yeshiva, known as Od Yosef Chai (Hebrew for "Joseph still lives," as spoken by Jacob after being told told that his son was alive in Egypt), was constructed at the site of the tomb, housing many sacred texts and the scroll of the Torah. An adjacent Israeli military outpost provided security for both students and visitors.

---

21 http://www.amana.co.il/?CategoryID=100&ArticleID=182.

Elon Moreh

It should be recognized that certain elements of the Israeli public are far from blameless when it comes to perpetrating hate crimes against Palestinian Arabs, some of which have been linked to the settlements in this area, including Yitzhar and the Od Yosef Chai yeshiva. The difference is that the IDF has cracked down on such vandalism (equally condemned by the settler leadership), whereas the Palestinian Authority celebrates the misdeeds of its own population.[22]

Under the 1993 Oslo Accords, Nablus, including Shechem and the Tomb of Joseph, were transferred to the Palestinian Authority. In September 1996, a cadre of rioters and gunmen stormed the tomb. Some reported that Palestinian police were involved, as the violence took the lives of six Israeli soldiers. Yeshiva students were among the many wounded. One student at the yeshiva named Hillel Lieberman recalled:

> After a large demonstration in the central square of Nablus, Arabs began to march on Joseph's Tomb. Within minutes, the tomb was surrounded on all sides by thousands of Arabs, who began to pelt the compound with rocks and firebombs. In addition, the same Arab policemen who regularly served on the joint patrol with the Israelis now began firing at the Israeli soldiers with their Kalashnikov machine

---

22 See Jacob Magid, "Yitzhar Settlement Viewed as Epicenter of Surge in 'Price Tag' Attacks," *The Times of Israel*, Apr. 23, 2018: https://www.timesofisrael.com/yitzhar-settlement-viewed-as-epicenter-of-surge-in-price-tag-attacks/.

guns.

Israeli soldiers [who were called as reinforcements] retreated into the thick-walled building which encompasses Joseph's tomb, and the Arabs advanced into the yeshiva [which neighbors the tomb], after setting fire to an army post and caravan which housed the Israeli soldiers. At this point, the Israeli soldiers again radioed for help.

The Arabs, upon entering the yeshiva, engaged in a pogrom reminiscent of scenes from Kristallnacht, when synagogues in Germany were ransacked by Nazis prior to the Holocaust. Sacred texts [in the yeshiva] were burned by Palestinians or torn to shreds, talises (Jewish prayer shawls) and sacred tefillin (phylacteries) were thrown to the ground, and anything of value was pillaged. The yeshiva's office equipment, refrigerators and freezers, beds, tables, and chairs, and all the food, were stolen.

The library was set afire, the structure of the yeshiva building was extensively damaged, and everything in it was in ruins. The only exception was the building of Joseph's Tomb itself, where the Israeli soldiers had taken cover, and the yeshiva's Torah scrolls, which the Israeli soldiers miraculously managed to save in the course of the riots.[23]

Israeli security forces ultimately managed to gain control of the situation, but no action was taken against the perpetrators, and one of the principal agitators, Jamal Tarawi, was in fact elected to the Palestinian parliament. Yasser Arafat called the whole incident a victory for the Palestinian resistance and a milestone in the effort to destroy the state of Israel. In the year 2000 Israeli Prime Minister Ehud Barak floated the idea of dismantling the military outpost as a peacemaking gesture, and instructed his security forces to prevent further Jewish visits to the tomb on designated days. In October of that year, following Arafat's return from the failed Camp David peace talks, a major terrorist offensive began, under the pretext that it was against the military outpost, not the tomb. A statement from the I.D.F., however, set the record straight:

The military compound set up next to the shrine was to safeguard and protect the site and the safety of worshipers, in accordance with the agreement with the Palestinian Authority. Claims by the Palestinians that the site was a military compound are blatant lies and possibly an attempt to legitimize the criminal and vulgar desecration of a Jewish

---

23　See Aaron Klein, op. cit., 90-91.

holy site.[24]

As events unfolded, Israeli border police were attacked by the al-Aqsa Martyrs Brigades, and a Jewish policeman was shot to death. Reinforcements were sent in.

Joseph's Tomb, "Welcome to Zion, Joseph the Righteous"

Mobocracy

After high-level negotiations in Paris, an agreement was reached by which Israel would pull back its troops in return for assurances that Palestinian violence would be halted. Although the Palestinian Authority promised that further damage to the site would be prevented, more violence broke out on October 5, when the funeral of one of the rioters was used as an excuse for unleashing a mob. Live fire from Palestinian gunmen was met in kind by Israeli security police. Prime Minister Barak, in the midst of relentless pressure, reached a decision on October 6 to withdraw entirely from the hallowed site, relinquishing it to the Palestinian Authority and the mob.

Before dawn on October 7, Israeli forces began an undercover evacuation of sacred objects, including a Torah scroll, from the tomb, only to be fired upon by Palestinians as they made their retreat. One border policeman was wounded in the incident, and some time thereafter, the site was completely overrun by rioters, who promptly hoisted a Palestinian flag over the edifice. Amin Maqbul, one of Arafat's spokesmen, declared,

---

24 Aaron Klein, *Schmoozing with Terrorists: From Hollywood to the Holy Land, Jihadists Reveal their Global Plans – to a Jew!* (Los Angeles: World Ahead Books, 2007), 41.

"Today was the first step to liberate al-Aqsa."[25]

The violence, however, was far from over. The entire contents of the structure, from furniture to holy books, were ravaged by the rioting horde. A reporter for the BBC observed: "The site was reduced to smoldering rubble – festooned with Palestinian and Islamic flags – by a cheering Arab crowd..."[26] A tragic footnote involved Hillel Lieberman, who had gone missing several days before, when he had set out to discover the nature of the unsavory events transpiring in the vicinity. On October 8, in a nearby cave, his body was found, having been shot repeatedly. On the charred remains of Joseph's Tomb and the yeshiva, a new mosque, with an Islamic green dome, was ultimately constructed.

One of the ringleaders of the mob, Abu Mujaheed, was quoted as saying:

> We used to attack the tomb on an almost daily basis. Our fighters were divided into five to six cells. Each cell was composed of ten members, and we used to crawl to the tomb and exchange fire with the soldiers. We led attacks in order to take away the Israeli flag. It was part of our resistance against the occupation. The goal was to chase the army of occupation from our land, and the result of [an Israeli retreat] proves that when we are determined we can defeat the Israeli enemy.[27]

He went on to declare that another yeshiva must never again be built in the vicinity:

> A yeshiva is an institution. An institution can be the beginning of claiming rights, and these claims can bring once again the Israeli army to establish a base in the place; and we cannot accept this. If the Jews try to build a yeshiva, we will shoot at them.[28]

Banned Again

As fate would have it, Jews were prohibited even from visiting the site until 2003. In 2006 Jewish students attempted to establish another yeshiva, constructing a building on Mt. Gerizim. They endeavored to locate themselves as close as possible to the remains of the beloved patriarch, so as to imbibe something of his eternal spirit. The main feature

---

25 Ibid., 39.

26 Ibid., 41

27 Klein, *Schmoozing with Terrorists*, 40; *The Late Great State of Israel*, 93.

28 Ibid., 94.

of the yeshiva was a synagogue, at which scores of students would regularly gather. Ironically, it was not the Palestinians who fomented trouble, but the government of Prime Minister Ehud Olmert. In the wake of bilateral peace talks with the Palestinians in the summer of 2007, Olmert had agreed to the dismantling of settlements deemed illegal. The new synagogue on Mt. Gerizim, which lacked an official government permit, was now classified as an "illegal outpost." As a consequence, Israel's own Prime Minister ordered the demolition of the synagogue on July 31, 2007.

View of Nablus from Mt. Gerizim

Assorted Jewish religious leaders within Israel came to the defense of the yeshiva students, noting that many illegal settlements have been erected across Judea and Samaria (the "West Bank") and Jerusalem by Palestinian Arabs, without being dismantled or in any way challenged by Israeli authorities. Indeed, hundreds of thousands of Arabs have come to take up residence in such locales. However, when it comes to religious Jews seeking to study the sacred texts in the vicinity of the reputed tomb of the patriarch Joseph, a cruel double standard instead provoked their forcible eviction. Jews are certainly accustomed to expulsion, having wandered across half the earth during their long centuries of torment at the hands of anti-Semitic princes and potentates. But to find themselves made refugees by their own government, being told in no uncertain terms that this area is to be *Judenrein*, was the bitterest of blows.

Even so, Palestinian militants were not finished desecrating Joseph's resting place. In early 2008 they set fire to sixteen tires, leaving them to burn in what remained of the tomb. The arson was later verified by the Israeli military, though Israel's own national media and news services chose not to cover the incident. Neither did American Jewish organizations or news outlets, possibly because they were unaware that it had even transpired.[29] A minor consolation may be found in the fact that an organization called Eshet Chayil (Hebrew for "Woman of Valor") funded the transference of a small number of sacred objects to Rachel's Tomb near Bethlehem.[30] They are safe there, at least for the time being, but in the ongoing geopolitical struggles over the future division of this "holy land," it is doubtful whether the impact of Israel's slow and tedious relinquishment of the Tomb of Joseph is fully grasped or comprehended.

A melancholy footnote may be offered, given that in recent years, the few Jewish visitors who chance to pay their respects at this third holiest site in Judaism have found the place a garbage-strewn ruin. Does Joseph "still live?" The one thing for certain, when it comes to this sacred locale, is that the battle for Joseph is far from over.

---

29 Klein, *The Late Great State of Israel*, 94-95.
30 Klein, *Schmoozing with Terrorists*, 44.

CHAPTER NINE

SURRENDERING HISTORY –
IN THE SHADOW OF SHILOH

ACCORDING TO THE Biblical record, Israel's great deliverer, Moses, was able to unite twelve semi-nomadic tribes into the worship of a single deity. He managed to lead them in what some call a concerted "revolt" from slavery into freedom. The exodus from Egypt amounted not only to a liberation from servitude; it was also the crucial event in the self-conception of the people called Israel. On a philosophical level, there was a powerful shift in thought when it came to monotheism. Pagan religions viewed human destiny as subject to the laws of nature, but at the foot of Mount Sinai, when Moses brought down the tables of the Law, human destiny was effectively separated from nature's vicious cycle. This people, Moses' people, finally broke free from fatalism and determinism. Life is progress, not repetition. This new "law of life" stressed the responsibility humans bear for their own destiny. What came forth from the mists of Sinai was a concept as revolutionary as the wheel, the idea that the human lot can be improved.[1]

The two tablets bearing the core of this revolution, the Ten Commandments, were deposited in a portable shrine, the Tabernacle of Moses, that accompanied the Israelites from place to place on their forty-year journey across the wilderness and toward the Promised Land. Unsurprisingly, the idea of the Tabernacle has been neither diminished nor erased from the Jewish psyche, in spite of the three and a half millennia since Moses commanded its construction. After so many centu-

---

1 See Abba Eban, *Heritage: Civilization and the Jews* (New York: Summit Books, 1984), 26.

ries, we may well ask: is tracking the Tabernacle through the occasionally trackless wasteland of biblical archaeology a proper quest? What are we to make of a growing modern "mania" surrounding the resting place of Moses' celebrated shrine? Has its very existence been irretrievably lost in the onward march of time, or does its final abode, presumably on a hilltop mound in biblical Samaria, undergird Jewish claims on the land itself?

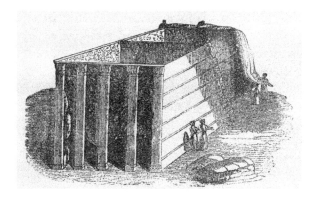

The Biblical Tabernacle

Tracking the Tabernacle

Palestinian Arabs routinely deny that Israelites ever occupied the ancient land called Canaan, maintaining that they, not the Jews, were its original inhabitants. Consequently, there are profound implications surrounding the archaeological evidence of the place biblically identified as Israel's great encampment in the Promised Land. Such evidence of course includes the site of the hallowed Tabernacle itself.

In an age when scholarly "consensus" finds it fashionable to consign hallowed tradition to the realm of sacred mythology, not a few contemporary academicians cast considerable doubt on the entire exodus narrative, many declaring that Israelites most likely never resided in Egypt at all. We are nonetheless told that under their great commander, Joshua, they set up camp, subsequent to their forty-year wilderness sojourn, first in a place called Gigal, and later in the important locale known as Shiloh. Situated some forty kilometers north of Jerusalem, it is described as "...north of Bethel, on the east side of the highway that goes up from Bethel to Shechem" (Judges 21:19). Beyond being a physical place, the very word "Shiloh" conjures up serious messianic imagery, in light of one of the most mysterious passages in the biblical text:

> The scepter shall not depart from Judah, nor the ruler's staff from be-
> tween his feet, as long as men come to Shiloh; and unto him shall the
> obedience of the peoples be. (Genesis 49:10)

Does this passage, in which the patriarch Jacob conveys his blessing
to the tribe of Judah, simply refer to all the tribes gathering at Shiloh as
a place of holy convocation, or to a future "anointed one," a "messiah," to
whom the people will flock? The latter option was the choice of the mas-
terful medieval Jewish commentator, Rashi, who declared that Shiloh is
a pseudonym for the Messiah himself and symbolic of redemption. In
any case, the importance of Shiloh among the Jewish people can hardly
be overestimated.

It was after the Israelite conquest of the land of Canaan that the city
of Shiloh was designated as a center of assembly for the twelve tribes. A
glaringly prominent passage in the book of Joshua declares:

> And the whole congregation of the children of Israel assembled them-
> selves together at Shiloh, and set up the tent of meeting there; and the
> land was subdued before them. (Joshua 18:1)

There could hardly be a detail more compelling than that relating
to the fabled "tent of meeting," consigned to a semi-permanent resting
place in the hilly sanctuary of Shiloh. The sacred tent was covered with
ten strips of fine linen, woven with blue, purple and crimson threads.
The text adds that the Tabernacle was to be covered by eleven cloths of
goat hair, fastened by loops and copper clasps. Ram and dolphin skins
were to be used as additional coverings. Beyond being a symbolic representation of the divine presence, it was said to have housed the most
revered object ever held by the Jewish people, the sacred Ark of the Covenant, a gold-covered chest adorned with two golden cherubim. In addition to the two stone tablets bearing the Ten Commandments, written
by "the finger of God," it contained the fabled rod of Moses' brother
Aaron, which miraculously budded to signify his authority, and a jar
of manna, the food which fell from heaven and which became the daily
staple of the wandering Israelites, as they trekked across the Sinai for
forty years.

The Ark rested in a chamber at one end of the Tabernacle, known
as the Holy of Holies, which could only be entered by the High Priest
on the Day of Atonement, at which time the blood of sacrificial animals was to be sprinkled on the sacred chest. At the portal of the Holy

of Holies was an intricately embroidered curtain, beyond which was a chamber containing a seven-branched menorah, the flame of which was constantly kept burning. Opposite the great candelabra was a table of "showbread," consumed by those performing priestly service. These sacred precincts lay in the center of an outer courtyard, enclosed by a rectangular fence, where a great altar provided a place of daily slaughter of sacrificial animals.

The Tabernacle was literally the place where the heavenly precincts intersected the "surly bonds of earth." This was not a house for the people; it was God's residence, and the focal point of worship for the entire nation. It was to this place that Israelites were said to have made pilgrimage for some three hundred sixty-nine years, until the Ark was captured by the Philistines. The irrepressible question is whether a location of such monumental significance as that of the Tabernacle has indeed been lost to time, or whether its shadowy traces may yet be discerned, archaeologically, in Shiloh.

The Telltale Tel

The place to look is the archaeological site known by the local Arabs as Khirbet Seilun (clearly a corruption of Shiloh), a layered mound, or tel, comprising an area of twelve acres, and rich with relics of the biblical era. It was the renowned American archaeologist, Edward Robinson, who, in 1838, first identified the location as biblical Shiloh. In 1922 a Danish expedition conducted the first excavation of the site, and in the same year the basic ceramic sequence at Shiloh was determined, thanks to William Foxwell Albright. Three excavation seasons ensued, conducted by another Danish team from 1926 to 1932. A series of soundings were conducted by the Danish in 1963, and between 1981 and 1984, Bar Ilan University's Israel Finkelstein conducted major excavations, revealing a major occupied city dating to the late twelfth century, or 1100s B.C.E. Finkelstein noted that Shiloh was destroyed not long afterwards, probably by the Philistines, a detail consistent with the biblical account, relating how the Ark of the Covenant was captured in the days when Eli was a judge in Israel. Locating Shiloh itself is something akin to an archaeological "slam-dunk," as the signs of its destruction are conclusively dated to the appropriate period, as referred to by the prophet Jeremiah: "Go now to My place that is in Shiloh, where I caused My Name to rest at first, and see what I did to it because of the wickedness of My people Israel" (Jeremiah 7:12). The site of the Tabernacle, however, would be a matter of considerable conjecture for the court of public opinion.

Ancient Shiloh

Exhibit A

The first to form a workable hypothesis as to the Tabernacle's location was Major Charles Wilson of the Palestine Exploration Fund. In 1866, Wilson concluded that the sacred shrine's most likely resting place would have been on a rocky escarpment situated one hundred forty-six meters to the north of the tel.[2] He supported the location with a simple comparison of the biblical dimensions of the Tabernacle enclosure and those of the scarp itself. Moreover, the bedrock had been deliberately flattened, as if to serve as a base for some sort of structure. Numerous holes had been dug at intervals into the bedrock, and while their exact date and purpose are a mystery, it is quite conceivable that they are related in some way to the Tabernacle itself. Muddling Wilson's conclusions was the fact that no remains from the time frame associated with the shrine, the early Iron Age, surfaced in the excavation. However, such remains did indeed appear in a later dig, essentially reopening the identification.

Exhibit B

The exact summit of the tel represents a second possibility, being supported by both Finkelstein and the Danish excavation. Finkelstein observed, "We can't rule out where the *mishkan* [Tabernacle] was, but if I say what's in my gut, I think maybe on the summit."[3] Indeed, there were sacred precincts located at the apex of many ancient sites, in the Near East and across Mediterranean lands, including the Acropolis of Athens. A large Crusader building at this location obscures much of the archaeologically rich terrain, but may preserve in its recesses important Bronze and Iron Age remains. An objection to this theory may be found in the fact that the summit area of the tel is by no means level. However, a perfectly level space is by no means a requirement, as evidenced by Jerusalem's Temple Mount, which was quite uneven until Herod constructed his massive artificial platform. Another objection is based on the biblical text itself, which declares, "You shall surely destroy all the places, wherein the nations that you are to dispossess served their gods, upon the high mountains, and upon the hills" (Deuteronomy 12:2). It therefore seems unlikely that Israel's most sacred shrine would be located at Shiloh's highest elevation.

Ancient Shiloh Summit

Exhibit C

A large, flattened plateau to the south of the tel is yet another pos-

---

3 Amanda Borschel-Dan, "With Bibles and Shovels, Search for Biblical Tabernacle Gathers Pace at Shiloh," *The Times of Israel*, July 17, 2017: https://www.timesofisrael.com/with-bibles-and-shovels-a-search-for-the-biblical-tabernacle-gathers-pace-at-shiloh/.

sibility. It appears likely that a pre-Christian sanctuary once stood here, later replaced by a number of Muslim shrines and several churches dating to the Byzantine era. A mosaic excavated in 2006 contains an inscription which reads: "Lord Jesus Christ, have mercy on Seilun [Shiloh] and its inhabitants, Amen." Clearly, the Christian residents from that period believed that they were at the place of biblical Shiloh. One of the Byzantine churches is specifically identified with the holy Tabernacle. At no other part of the ancient site was there such a religiously dedicated construction. One final option is that the Tabernacle was originally placed at the summit of the site, but relocated to either the northern or southern locations.

Obviously, the excavations are of significance, not only to Jews, but to Christians as well. Professor Scott Stripling of the Associates for Biblical Research observed, "A sense of awe came upon me as I contemplated how God had set before us an open door at Shiloh which will have a direct impact on how people read their Bibles in the future."[4] Among the objects revealed was an Iron Age four-horned altar, reused in a Byzantine wall, where it lay unrecognized until 2013. Its presence at the site bears witness to animal sacrifices being practiced at ancient Shiloh.

Former Arkansas governor, Mike Huckabee, a Baptist minister and evangelical Christian, referred to the place as Israel's first capital and declared, "Shiloh is proof from three thousand years ago this land was home to @Israel site of ancient Tabernacle."[5] Today the site is under the management of a local council of Israeli settlers, known as Mishkan Shiloh. With the support of the Israeli government, which has funneled over four million dollars into upgrading the site and a new visitors' center, tourism has increased exponentially to over 120,000 visitors per year. The visual effects are impressive, including a holographic representation of what the ancient shrine may have resembled. A large majority are evangelical Christians, magnetically drawn to the tel, now known as Shiloh: City of the Tabernacle.[6]

Set-tel-ers

For the Israeli settlers the interest is welcome, and a boon to their own determination to remain in this place, living, as it were, on top of the Bible. The settlement was established, like so many others, in the

---

4 Ibid.

5 Ilan Ben Zion, "Ancient West Bank Site Draws Christians, and Controversy," *AP News*, March 27, 2019: https://apnews.com/a8bfb603733f4cb3890bbb4fcecd06c1.

6 Ibid.

wake of Israel's surprise victory in the Six-Day War. Jews wanting to exercise their personal "right of return" to the biblical Land of Promise initially turned their gaze toward Shechem, and the Palestinian Arab city of Nablus. Their plans having been foiled by the Israeli Defense Force, who physically ejected them, they began moving in on the area around Shiloh, under the "cover" of an archaeological excavation.

The plan was the brainchild of Ira Rappaport, a member of the same "Jewish Underground" that had perpetrated terrorist acts in Hebron, Nablus and other places, targeting Palestinian mayors and murdering Muslim students. Found guilty of detonating a car bomb that blew off the legs of two Palestinian officials, he became leader of the Shiloh community after serving only one year in prison. As noted, however, Israel does punish its homegrown terrorists, whereas the Palestinian Authority most often extols them. Rappaport's daughter commented:

> My father was very connected to Israel since he was a young child. He volunteered in a hospital during the Six-Day War. His whole feeling was that he was coming back home, and he was very passionate.[7]

An inscription on the grave of an Israeli mother of seven, murdered on her way to Tel Aviv by a Palestinian terrorist, reads: "Here in the bend of the road, in the fold of the mountain, we swear that the covenant will never be broken."[8] Spurred on by such determination, the settlers managed to establish a series of outposts in the valley of Shiloh, where more than four thousand Israelis now live, notwithstanding opposition from the international community and the Israeli government itself. Asserting their right to return to their ancient homeland, their "revenant rights," the community is served by shops, educational institutions, and a synagogue, called Mishkan Shiloh, built to resemble the biblical Tabernacle. The Hesder Shiloh Yeshiva was opened in 1979, with over one hundred fifty rabbinical students who spend their days pouring over sacred texts. Regrettably, some acts of vandalism against nearby Palestinian inhabitants have been perpetrated by certain settlers, though it should be noted in fairness that the great majority are peaceful, viewing life in the area as an opportunity for improved quality of life, in what amounts to a "bedroom community" of Jerusalem.

---

7 Jake Wallis Simmons, *Meet the Settlers: A Journey through the West Bank*, Chap. 3: "The Battle for the Hilltops," *The Telegraph*: https://www.telegraph.co.uk/meetthesettlers/chapter3.html.

8 Ibid.

Mishkan Shiloh Synagogue

Emek Shaveh and another Israeli NGO, Yesh Din, unsurprisingly raise their voices in strident complaint, charging that the settlers are attempting to "…reinforce the connection between the biblical Shiloh and the modern settlement, in a manner not necessarily based on the archaeological discoveries at the site." They allege that the goal of the archaeological Park is to "…create a broad consensus about its importance as an indivisible part of the state of Israel."[9] It is considered politically incorrect to recognize what lay behind such criticism – a concession to the victimhood status of the "suffering Palestinians." Nevertheless, there are indeed implications to archaeology, and for many it is difficult to comprehend how those recognizing the eternal importance of a place like Shiloh to the Jewish people should be castigated as insensitive.

By way of historical comparison, it might be noted that in the middle of the nineteenth century, the United States essentially "conquered" Texas and California in the Mexican War, annexing the former, purchasing the latter, and repopulating these areas with Americans of European descent. Today, there are not a few politically correct voices justifying what may one day amount to a slow annexation of these "lost provinces" by Mexico.

Millions of Mexican immigrants, both legal and illegal, are creating their own "facts on the ground" to further the cause. Where are the voices of indignation on the left, condemning such "illegal settlers?" Instead, it is argued that America's borders should be opened wide to illegal Mexican immigrants. However, when it comes to Israeli Jews seeking to settle in land seized from the Jewish people long ago, by one

---

9   Ben Zion, op. cit.

conquering army after another, it is a different story. Never mind that Arabs, over several centuries, moved into these territories, which had formerly been Jewish, and that these Jewish settlers want them back. In this case, it is the Palestinians who are "victims" and the Jews who are usurpers – usurping the land of their ancestors. It might well be asked what will become of these settlers should a Palestinian state become the effective sovereign over Judea and Samaria. Doubtless, they will not be tolerated. They will be forced to abandon their homes, their possessions, the entire infrastructure they have labored to build. They will become history's latest cohort of homeless, evicted, and wandering Jews.

Settling in Samaria, Dwelling in Dothan

Shiloh is only one example of multiple attempts to establish Jewish settlements in the biblical heartland. A prime example is the ancient city of Samaria. At one time the capital city of the northern kingdom of Israel, it is one of the most impressive archaeological sites in the entire region. In late 1976, members of the now-defunct Gush Emunim ("Block of the Faithful") settlers movement attempted to establish a community here, only to be evicted by the Israeli government.

Not far from Shiloh, the biblical city of Dothan is yet another magnet for Israeli settlers, unsurprisingly drawn to a place immortalized in the story of Joseph and his treacherous brothers. Located twenty-two kilometers north of Shechem, and about ten kilometers southwest of the Palestinian Arab town of Jenin, it was here, about three millennia ago, that the favorite son of Jacob was betrayed by his jealous siblings. The eldest of the twelve, Rueben, persuaded the others to spare the lad, whereupon they decided, rather than shedding Joseph's blood themselves, to throw him into a pit. Later they sold him into slavery to a passing caravan of Ishmaelites. Dothan is located on an ancient road, which winds its way from the region of Gilead in the north, down to the land of Egypt. This detail perfectly suits the narrative, which relates that Joseph is next taken as a slave to the land of the pharaohs.

On a linguistic level, the Hebrew word for pit happens to be "dot," which is the root behind the word Dothan. Indeed, this ancient city is well known for the many ancient pits in its immediate vicinity, one of which is particularly deep and an excellent candidate for the one made famous by the Joseph story. The director of the Samaria Study Center, Yair Elmakayes, commented:

> We are at the foot of the city of Dothan, which has been identified for certain, and its excavation reports were already published in 2005. To-

day there is no question about it – this is the Biblical Dothan.[10]

As is so often the case, the Arabic place name, Jubb Yussef ("Joseph's Well"), uncannily recalls the essence of the biblical site. Elmakayes continued:

I am one hundred percent sure that the impressive hill in front of us is the Biblical city of Dothan. Pointing at a certain place and saying that a Biblical event happened here 3,500 years ago is extremely problematic. But there is the connection of the city of Dothan with wells next to it. It wouldn't be unfounded to say that if the Biblical story did happen, it happened here.[11]

Tel Dotan

The excavated site of Tel Dotan is perched on a natural hill, roughly forty-five meters in elevation, but the layer-cake remains of ancient habitation bring the total height to sixty meters, affording an unobstructed view over the surrounding valley. The summit area consists of some ten acres, and the steeply inclined flanks add another fifteen acres to the

10 Assaf Kamar, "In search of Joseph's well: A journey into the Book of Genesis," *Ynet News*, Oct. 4, 2016: https://www.ynetnews.com/articles/0,7340,L-4860333,00.html.
11 Ibid.

total.[12] Eight seasons of excavation in the 1950s revealed Dothan as one of the most ancient cities in the Middle East, its earliest remnants dating back fully five thousand years, to roughly 3000 B.C.E. As the rubble was cleared, houses, walls and streets came into view, peppered with the remnants of pottery and household utensils. From such evidence it is clear that Dothan was, by and large, continuously occupied until around 700 B.C.E., when, as revealed by a layer of destruction, it met its final demise.

Archaeological remains dating from the days of the patriarchs, roughly 2000–1800 B.C.E., were found in the debris of two of the lower levels. Consequently, even if the story of Joseph is a fanciful concoction, it was certainly accurate with regard to the existence of the city of Dothan, in the right place and at the right time, namely 1900-1800 B.C.E. Dothan is further referenced by the biblical narrative, when the king of Aram is said to have besieged the city in order to apprehend the Prophet Elisha (2 Kings 6:13). While the details of the story seem incredible enough (the prophet smites a host of his adversaries with blindness and leads them to the city of Samaria), there is no question that Dothan existed at this time as well (850–800 B.C.E.), based on findings from the tel's upper layers. Archaeology by itself can never confirm the Bible, but in the case of this ancient site, it certainly does not hurt.[13]

Aware of the significance of the place, a dedicated corps of Israeli settlers set up tents nearby, in October 1977, calling their fledgling community Mevo Dotan ("Approach to Dothan"). In 1981 they moved to a barren hilltop in the vicinity, known as Jabl al-Akra – "Bald Mountain." It essentially amounted to exerting Jewish "squatters' rights" on unoccupied land owned by the state. However, with the outbreak of the Second Intifada in 2001, Jewish Israelis became moving targets. Faced with rapidly deteriorating security, the majority of the settlers simply moved away. Two years later, people began returning to the settlement's abandoned homes, with the founding of an institute for the study of the Talmud and rabbinic literature. Nonetheless, as of 2014, the Jewish population of the town was a scant three hundred thirty.

---

12 http://ancientneareast.tripod.com/Dothan.html

13 See Joseph P. Free and Howard Frederic Vos, *Archaeology and Bible History* (Grand Rapids: Zondervan, 1992), 69. See also Joseph P. Free, "The First Season of Excavation at Dothan," *Bulletin of the American Schools of Oriental Research* 131 (Oct. 1953): 16-20; "The Third Season at Dothan," *BASOR* 139 (Oct. 1955): 3-9; "Digging Down to Ancient Dothan," *Moody Monthly* (Nov. 1954): 15-17ff.

Mevo Dotan

In line with the Disengagement Agreement of 2005 four Israeli settlements in the area (Sa-Nur, Homesh, Ganim and Kadim) were dismantled, their inhabitants forced to relocate. To ask what Arab government would forcibly resettle its own citizens would be an absurdity; yet,
the state of Israel did exactly this – again and again. The residents of
Mevo Dotan were understandably fearful that they would be next. Fate
nonetheless intervened, and, during the Passover feast of 2016, a new
housing project was inaugurated, projected to double the population of
the community.

The Israeli public has long been divided on the wisdom of establishing and maintaining the settlements, but Knesset Speaker and Likkud
MP, Yuli Edelstein, hailed it, appealing to the immortal line from Psalm
118:22: "The stone which the builders rejected has become the chief cornerstone." To this he appended a prayer of his own: "May Mevo Dotan
and all the settlements in the area be the cornerstone for the growth of
northern Samaria."[14] Nevertheless, one must still ask whether, with the
potential creation of a Palestinian state, the "cornerstone," and the settlers who laid it, will be rejected once more.

Searching for Samuel

One of the most revered figures of Jewish tradition is the great

---

14   Tovah Lazaroff, "Settlers Celebrate New Housing Project to Double Size of Mevo
Dotan Settlement," *Jerusalem Post*, Apr. 27, 2016: https://www.jpost.com/Arab-Israe-
li-Conflict/Settlers-celebrate-construction-project-in-Samaria-452442.

prophet of the early Israelite confederacy, Samuel, who anointed the nation's first king, Saul, and later, the mighty King David. Today, perched on a high place, some nine hundred meters above sea level, and overlooking the hill country of Judea, is Samuel's reputed tomb. As with other biblical sites, it is strategically located, in this case guarding the important trade routes that connected the ancient cities of Emmaus, Bet Horon and Jerusalem.

Nabi Samwil, early 1900s

The prophet's name is preserved in the Arabic designation of a mosque built above a Crusader fortress, covering the burial site – Nabi Samwil. But was the great prophet indeed buried here? The scriptural account reads: "And Samuel died; and all Israel gathered themselves together, and lamented him, and buried him in his house at Ramah" (1 Samuel 25:1). Ramah in Hebrew refers to a high place, but of all the high places in the land of Israel, why should this be the prime candidate? There are in fact other possibilities, such as a village called "A-Ram," to the northeast of Nabi Samwil, and Ramallah, due north. Even if Samuel had not been buried here, it has been suggested that his bones were reinterred at this location. It has also been argued that Samuel's supposed tomb is simply a shrine in his honor.[15]

Nabi Samwil has also been connected with the biblical location "Mizpeh" (Hebrew for "tower"), where the prophet was said to have presented the newly anointed Saul to the assembled Israelites:

> And Samuel called the people together to the LORD at Mizpeh... So Samuel brought all the tribes of Israel near, and the tribe of Benjamin was taken... And he brought the tribe of Benjamin near by their families... And Saul the son of Kish was taken... And Samuel said to all the people: "See him whom the LORD has chosen, that there is none like him among all the people?" And all the people shouted, and said: "Long live the king" (1 Samuel 10:17, 20, 21, 24).

Some theorize that the linkage of this hill with the biblical Mizpeh forever connected it with Samuel, conferring on it an aura of holiness and ultimately identifying it as the prophet's burial place as well.[16] One more possible link is with the venerable King Solomon, who beseeched God, not for riches or power, but for wisdom:

> And the king went to Gibeon to sacrifice there; for that was the great high place; Solomon offered a thousand burnt-offerings upon that altar. The LORD appeared to Solomon in Gibeon in a dream by night; and God said: "Ask what I should give you." And Solomon said: ... "Give your servant therefore an understanding heart to judge your people, that I may discern between good and evil." (1 Kings 3:4-6, 9)

Might this "high place," identified here as Gibeon, be the same as what came to be called Nabi Samwil? Israeli archaeologist Yitzhak Magen excavated the site during several seasons, between 1992 and 2003, finding, on the steep southeastern embankment, remains dating back to the eighth and seventh centuries B.C.E. Some tombs in the area were found to date from the First Temple period, bolstering the possibility that the tomb of Samuel may have been here. Magen believed that he had identified the "Mizpeh of Benjamin" linked to the prophet Samuel. He acknowledged, however, the limitations of his own findings:

> We did not find any remains from the time of the Judges... not a single structure or even a standing wall from this period. On this basis, it might be tempting to conclude that the site was unoccupied at this

---

15 https://www.biblewalks.com/nebisamuel
16 Ibid.

time... [17]

For centuries Jews have made regular pilgrimage to the site, to pray in an underground chamber housing a small synagogue, all the while respecting the mosque that was also built there. The late nineteenth century saw Jewish efforts to establish a village at Nabi Samwil, which they called Ramah, in honor of the prophet Samuel. Land in the vicinity was legally purchased, on which thirteen Jewish families of Yemenite extraction have settled, at least for the time being.

Over the last two millennia, throughout the Common Era, it remained strategically important, due to its elevation alone. There are abundant remains, including fortifications, spanning multiple centuries, including the Roman-Byzantine era (attested by a large monastery), the Crusader period, the Arab-Mamluk period, and the Ottoman period. Heavy fighting between the British and the Turks in World War I resulted in the destruction of the village dating to the Ottoman and Crusader periods. The war's end saw the restoration of the mosque.

During Israel's 1948 War of Independence, Arab forces used the site to bombard Jewish convoys, traveling up the steeply inclined road from Tel Aviv in a desperate attempt to resupply besieged Jerusalem. Israeli troops attempted, but failed, to overrun the site, suffering severe casualties. From 1948 to 1967, the site was taken over by the Arab Legion, which used it as a military post to guard Arab East Jerusalem. The land purchased by Jews in the previous century was confiscated, with neither outcry nor formal condemnation from the international community.

It was not until the Six-Day War, when Nabi Samwil was again used as a base for Arab attacks on Jews that it fell to Israeli control. The site's significance was not lost on Jewish settlers, who built a number of homes there. In 1971, Israel itself demolished them. Eventually, Israelis did settle the area, effectively turning it into a suburb of Jerusalem. A national park was established around the ancient shrine, drawing tourists in large numbers and adding to the region's prosperity. Nonetheless, Israel is charged with uprooting the Arab inhabitants and demolishing the Arab village around the mosque, notwithstanding that the majority of its occupants had voluntarily fled during the war of 1967. It should be noted that the Israeli government repeatedly pleaded with the Arabs across the contested lands, urging them not to flee and insisting that their rights would be respected. They chose instead to believe the Arab

---

17 See Yitzhak Magen, "Nebi Samwil: Where Samuel Crowned Israel's First King," *BAR* 34/3 (2008): 36-45, 78-79.

propaganda mills, which insisted that they would all be massacred.

A cacophony of condemnation continues to this day, from all of the usual, politically correct suspects. Palestinians are "living inside a cage,"[18] so it is claimed. The "separation barrier" Israel constructed to prevent Palestinian terrorists from entering Israel proper has isolated the Arabs of Nabi Samwil in what activists call an "invisible cage." But such voices of rage make no mention of the initial reason for the barrier's construction – to prevent the rash of suicide bus bombings against in-nocent Israelis, traveling on public transportation. It is radical elements of the Palestinian Arab population who are waging an endless war of attrition against Israel, and in light of the historical use of Nabi Samwil as a base for the ruthless bombardment of their Jewish neighbors, it is no small wonder that the Israeli side would unofficially "annex" it, say-ing: Enough!

Nabi Samwil today

There remains no shortage of voices clamoring for Palestinian rights, but where are the advocates for Jewish rights? Moreover, if and when a Palestinian state is established, and if it is based on the 1949 armistice lines, will Nabi Samwil revert to Arab sovereignty? If so, what will become of the Israelis who now live there? What will become of their homes and their property? How many voices of condemnation will he sounded at the inevitable confiscation of Israeli property by the Arab side? The one thing that can be said about the Israelis is that this time, unlike their long history of foreign conquest and dispersion, they will not likely surrender easily.

Remembering Rachel

There is one more ancient tomb that must be mentioned, that of the biblical matriarch Rachel, which was destined to enter the crosshairs of Middle East turmoil. Located on the northern outskirts of the city of Bethlehem, it is considered holy to Jews, Christians and Muslims. It is also a symbol of the Jews' return to their biblical homeland.

Rachel's Tomb, 1930s

According to biblical tradition, Rachel was the beloved wife of Jacob, for whom he labored fourteen long years. She died in childbirth, while en route to Hebron, only a short distance from Bethlehem. Jacob buried her by the side of the road. Many centuries later, the Israelites would be driven into captivity in Babylon, trudging along this same road and deriving courage and comfort from Rachel's mystical presence. The prophet Jeremiah poetically immortalizes their experience:

A voice is heard on high,
Lamentation, bitter weeping,
Rachel weeping for her children,
She refuses to be comforted for her children
For they are not. (Jeremiah 31:14)

While there is no archaeological evidence that Rachel was actually buried here, the site, marked by a diminutive dome supported by four beams, dates back to the fifth century of the Common Era, and has been revered as authentic. In the nineteenth century the prominent British philanthropist, Sir Moses Montefiore, while traveling across the holy land, saw to the construction of walls and a visitation room for pilgrims.

Rachel's Tomb in 1978 and 2018

Following Israel's War of Independence in 1948-1949, Bethlehem and the surrounding territory was annexed by Jordan, after which Jews no longer had access to the site and could no longer pray there. A Muslim cemetery was planted, surrounding the tomb, and Bethlehem grew to such an extent that Rachel's resting place was now in the middle of the town. Following the Six-Day War, Rachel's tomb was again opened to Jews, who frequently came singing: "Your sons have come back to you, Mother Rachel, at their head Benjamin and Joseph... We will never go away from here again, Rachel."[19]

Unfortunately, things deteriorated following the Oslo Accords of 1993. After Israel ceded control of the area to the Palestinian Authority, the area could no longer be adequately protected. In 1996, following the opening of Jerusalem's Western Wall Tunnel, a rioting Arab mob began descending on Rachel's Tomb, intent on its complete destruction. To their credit the Palestinian police intervened, forming a human chain as one of their number shouted over a bull horn, "Enough bloodshed; please listen to us!" Though taunted and stoned by the crowd, it was a rare moment of sanity, resulting, several hours later, in the crowd's dispersement and heartfelt thanks from an Israeli officer, who shouted, "We appreciate what you have done here!"

The Jewish state, in response, constructed a small fortress around the tomb, including guard towers, barbed wire and concrete walls. The construction workers had to cope with Arab rioters, armed with weapons. Is this the care and protection that a future Palestinian state might afford to Jews who wish to visit the holy sites and archaeological remains of an ancient Jewish civilization?

In 2005, the tomb was enclosed on the Israeli side of the "separation barrier." Since 1996, Muslims refer to it as the site of a mosque built at the time of the Arab conquest of 632 C.E. UNESCO chimed in as well, declaring Rachel's Tomb as the Bilal ibn Rabah Mosque. The site's Jewish heritage is denied outright. Israel was called upon to rescind its declaration of Rachel's Tomb and the Tomb of the Patriarchs in Hebron as national heritage sites. The strong protest registered by Israel's ambassador to UNESCO was officially expunged from the record.[20] One cannot help but wonder, given the rising tide of anti-Semitism internationally, whether the world community would just as easily seek to "expunge" the Jewish people themselves.

---

19 Nechama Golding, "Rachel's Tomb (*Kever Rachel*)" *Chabad.org*: https://www.chabad.org/library/article_cdo/aid/602502/jewish/Rachels-Tomb-Kever-Rachel.htm.

20 Nadav Shragai, "Until 1996, Nobody Called Rachel's Tomb a Mosque," *Jerusalem Post*, Nov. 8, 2010: https://www.jpost.com/Israel/Until-1996-nobody-called-Rachels-Tomb-a-mosque.

CHAPTER TEN

"LOOTING" HISTORY – DEAD SEA SQUABBLES

THE WORLD CHANGES, silently and without fanfare, on a particular summer day in 1947. A goat belonging to a Bedouin flock wanders off among the multiple caves that dot the desert cliffs near an ancient ruin known in Arabic as Khirbet Qumran. Young Muhammad Adh-Dhib is a Bedouin lad, responsible for tending these goats, and it should come as no surprise that he becomes more than a little agitated over the fate of this lost animal. He runs off to search among the rocky crags of the Judean wilderness. After a time he begins hurling stones into the cave entrances, hoping to frighten any wayward beast out into the open.

Suddenly, young Muhammad lets loose a stone that produces another sound, the sound of shattering earthenware. The stone has broken open a large clay vessel nestled along the cave wall. The Bedouin lad pulls himself up the cliff and peers into the cave entrance. He spots several tall earthen jars, with broken pottery fragments laying about. His first thought is that this might be the lair of ghosts and desert demons. He darts away from the cave, returning to his Bedouin camp. The next day, Muhammad and a friend return to the cave. Summoning their courage, they make their way through the small opening and begin opening the jars, only to find that most are empty. Peering into one, however, their eyes are met by three odd-looking bundles of flaxen-covered, rolled leathery vellum parchments.

Unbeknownst to them, they have stumbled upon one of the most important archaeological finds of the twentieth century, which came to be known as the Dead Sea Scrolls. Equally unknowable was the degree to which even this treasure-trove of biblical and sectarian documents, composed prior to the days of King Herod the Great, Jesus of Nazareth and John the Baptist, would become the focal point of yet another international dispute, involving Israel and the Palestinian Arabs. The most important fact of all is that these are Hebrew documents which predate Christianity, let alone Islam. They are incontrovertible evidence of a Jewish presence in the land before the Common Era.

Dead Sea Scrolls jars

It should come as no surprise that one of the demands of the Palestinian Authority is for Israel to "return" and "repatriate" them to their own nascent state. The claim is that the Dead Sea Scrolls amount to "looted" Palestinian artifacts. Indeed the Scrolls were discovered, not on Israeli territory, but on land later claimed for "Palestine." Are such arguments serious, or properly dismissed as delusional? Are the world's oldest remnants of the Hebrew Bible the rightful possession, not of the Jewish people, but of Muslim Arabs? Certainly, a modicum of additional backstory is in order.

Bedouin Banditry

It is of course illegal to remove antiquities from their place of discovery, and from the moment young Muhammad and his friend pried open the jar containing the scrolls, they were engaging in a bit of banditry. Upon extracting the fragile parchments from the ancient pot, they returned to their Bedouin campsite and began to unroll their dubious find. The soft, leathery material was covered with curious-looking lettering which they could not read. The script was ancient Hebrew; their own tongue was, of course, Arabic. Puzzled, they decided to carry the bundles around with them for a number of days. During the weeks that followed, the scrolls were tucked in a bag dangling from a tent pole. Finally, the decision was made to take them to the open market (in Arabic, the '*souk*') in Bethlehem, where they brought them to a local shopkeeper, nicknamed Kando, who ran a small general store and cobbler shop.

Kando purchased several of the parchments, but his native tongue was, likewise, Arabic, and he was no better suited to read the lettering than the Bedouin. For some time, he wondered whether the leathery vellum might, if cut into strips, be suitable for fashioning sandal straps. However, on examining the letters again, he decided that the parchments might indeed be valuable. Kando and an accomplice decided to return to the vicinity of the cave where Muhammad had encountered the scrolls and began searching the area. Indeed, they discovered several other jumbled wads of vellum, along with numerous fragments, which they now recognized as potentially precious artifacts. Kando's next move was to take four of his scrolls to the Old City of Jerusalem, to show them to the Christian elders of the Syrian Orthodox Church, to which he belonged. His destination was St. Mark's Monastery, in the Armenian Quarter of the city.

The head of the monastery, the Metropolitan Samuel, was a Syrian Orthodox cleric, whose handsome visage and priestly robes conveyed an almost regal bearing.[1] The Metropolitan was clearly intrigued by the strange documents, but was equally unable to read the Hebrew lettering. His hunch was that the writing might be ancient Syriac. In any case, he eagerly bought them for the sum of twenty-four Jordanian pounds, which at the time was the rough equivalent of one hundred U.S. dollars.

In the meantime, Kando visited an antiquities dealer of Turkish-Armenian ancestry, whose shop was located in the crooked streets of Jerusalem's Old City. Kando sold him a few more of his parchment frag-

---

1   See Edmund Wilson, *The Scrolls from the Dead Sea* (London: W. H. Allen, 1955), 12.

ments. The following day, November 23, 1947, the Armenian placed a phone call to the famed archaeologist of the Hebrew University of Jerusalem, E. L. Sukenik, and arranged a secret meeting on the following morning, at the barbed wire dividing Arab East Jerusalem from Jewish West Jerusalem. At last, a Hebrew-speaking modern Israeli had a first look at a Dead Sea Scroll fragment, and could actually read and understand the ancient writing. Gazing at a single scrap of parchment, he was amazed. On Nov. 25, 1947, he wrote in his journal:

> Today I met the antiquities dealer. A Hebrew book has been discovered in a jar. He showed me a fragment written on parchment.[2]

Sukenik decided that he needed to see more of these parchments, and he determined to travel directly to the source, Kando's shop in Bethlehem. Complicating the situation, Bethlehem was located in Arab territory, in what would later become part of the Hashemite Kingdom of Jordan. Sukenik lived in Jewish Jerusalem, in what would in a matter of months be reborn, amidst war and bloodshed, as the new capital of the State of Israel. Such realities were destined to exacerbate the controversy of who rightfully owns the Dead Sea Scrolls.

Crossing Borders
Travel between the two areas was risky, but Sukenik found it necessary to make the clandestine journey. What he discovered in Kando's shop far exceeded his expectations. His journal entry that day declared:

> I saw four pieces of leather with Hebrew writing. The script seems ancient to me… Is it possible? He says there are also jars. I looked a bit and found good biblical Hebrew, a text unknown to me. He says a Bedouin of the Ta'amireh tribe brought it to him.[3]

He later wrote: "It may be that this is one of the greatest finds ever made in Palestine, a find we never so much as hoped for."[4] Kando relinquished three of his scrolls, which he had not sold to the Metropolitan Samuel, to Professor Sukenik, who quickly became aware of the aston-

---

2 James VanderKam, Peter Flint, *The Meaning of the Dead Sea Scrolls: Their Significance For Understanding the Bible, Judaism, Jesus, and Christianity* (London: T&T Clark, 2002), 6.

3 Wilson, op. cit., 20.

4 VanderKam, op. cit., 7.

ishing documents he had acquired:

- a partial text of the book of Isaiah, written by the hand of a Jewish scribe, two centuries before the Common Era,
- a collection of non-biblical Psalms, which resemble Psalms from the Bible, but which had never been seen by modern eyes until 1947, and
- a strange document describing an apocalyptic battle, the War Scroll.

The War Scroll

It should be noted that however the scrolls were removed, or "looted," from their original location, in desert caves, they had now been legally purchased, by a Jewish Israeli, on behalf of the Hebrew University. Sukenik recalled:

> My hands shook as I started to unwrap one of them. I read a few sentences. It was written in beautiful biblical Hebrew. The language was like that of the Psalms, but the text was unknown to me. I looked and looked, and I suddenly had the feeling that I was privileged by destiny to gaze upon a Hebrew Scroll which had not been read for more than two thousand years.[5]

An incredible footnote of history is the fact that on that very evening, November 29, 1947, the United Nations voted on the partition of British-controlled Palestine, paving the way for the creation of the modern State of Israel. Sukenik's journal entry reads:

> While I was examining these precious documents in my study, the late news on the radio announced that the United Nations would be voting on the resolution that night – whether or not Israel would be allowed to become a nation – My youngest son Mati, was in the next room, twiddling radio knobs in an effort to get New York... It was past midnight when the voting was announced. And I was engrossed in a particularly absorbing passage in one of the scrolls when my son rushed in with the shout that the vote on the Jewish State had passed. This great event in Jewish history was thus combined in my home in Jerusalem with another event, no less historic, the one political, and the other cultural.[6]

Nothing could more effectively underscore the inextricable linkage between archaeology and geopolitics in today's Middle East. For Sukenik the recovery of this ancient library was nothing less than a "sign," lending a kind of divine legitimacy to the yet unborn Jewish nation. Back in Bethlehem, Kando had not relinquished all of the scrolls in his possession. There were several large scroll fragments with which he was unwilling to part. Moreover, he feared that he would be held responsible for removing antiquities from a historical site, a crime punishable under both Jordanian and Israeli law. For his own protection, he decided to bury these large parchment fragments in his garden. Unfortunately, when he later dug up the spot, in order to retrieve them, he found, instead of carefully inscribed scrolls, a tarry, mangled muck. We can only speculate on what those parchments might have contained.

Later, in Jerusalem, at the end of January 1948, the Metropolitan Samuel directed an associate write a letter to none other than E. L. Sukenik, asking his judgment on some old parchments and making a fictitious offer to sell them. They arranged to meet in the Arab sector of Jerusalem, which was still divided by political turmoil. Sukenik used the occasion to "borrow" the four mysterious documents, which he took home to examine. These parchments, by all appearances, were even more valuable than his other three. He identified their contents:

- another scroll of the book of Isaiah, some twenty-four feet in length, virtually entire and complete, containing all sixty-six chapters;

---

6 Quoted by Yigael Yadin from the diary of his father, Eliezer Sukenik, in *Yigael Yadin, The Message of the Scrolls* (New York: Crossroad, 1992), 24.

- a previously unknown commentary on the small and often over-looked biblical prophet, Habakkuk (the Habakkuk Commentary), unknown until this moment;
- a manual of rules for membership in an exclusive Jewish religious order (the Manual of Discipline), most likely the sect which produced the scrolls;
- a clever reworking of the book of Genesis (the so-called Genesis Apocryphon), containing many previously unknown expansions on the biblical account.

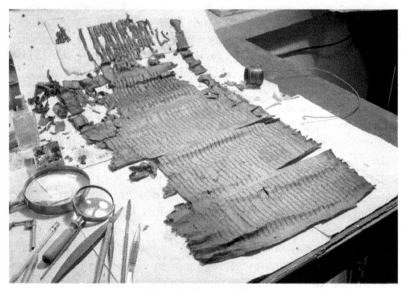

Genesis Apocryphon

Of course, Sukenik eagerly offered to buy them, even though funding was not available. When they met again, he reluctantly returned them to the Metropolitan's associate. He would never see those four parchments again. Undaunted, Professor Sukenik now set to work transcribing and editing the three scrolls remaining in his possession. He published the first of the documents between 1948 and 1950. Surprisingly, the scrolls were not instantly heralded as a major find. Many in the academic community (such as the renowned scholar, Solomon Zeitlin) failed even to believe that the scrolls were authentic. It was charged instead that they were medieval forgeries, since, it was reasoned, documents that old simply could not survive in caves for two thousand years. They must have been forged.

Other scholars nonetheless rallied to Sukenik, convinced that these do in fact date from the days of ancient Judea, in the first two centuries, B.C.E. From the standpoint of paleography, the Hebrew letters resemble those found on tombs and other stone monuments from the same period. Furthermore, the scrolls were eventually subjected to radiocarbon-14 dating. Such testing was not without its challenges. Since an unacceptably large amount of scroll material would have to be destroyed in this testing, the decision was made to test the scroll covers, the flaxen material enshrouding the parchments themselves, which the Bedouin had pulled away to reveal their treasure.

The tests were conducted by a scholar from of the University of Chicago, who established that the material is fully two millennia old. The date arrived at for the flax was 33 C.E., plus or minus two hundred years. Therefore, the scroll covers may date from as early as 168 B.C.E. and no later than 233 C.E. Of course, the scrolls themselves might be even older than the flaxen covers meant to protect them. Combined with the compelling evidence of handwriting analysis, most scholars have come to the favor the earlier part of this time frame as the likely period of composition of the majority of the scrolls, the second century, B.C.E. Thus, it was conclusively demonstrated that the scrolls are both authentic and Jewish.

Coming to America; Back to Qumran

For his part, the Metropolitan Samuel had other plans for his treasures. He arranged to bring all four of his scrolls to the United States, revealing them, amid great fanfare, at a special ceremony in the Library of Congress. Rumor had it that he hoped to sell them for a reported sum of at least one million dollars. His asking price was well beyond the reach of even the most serious investors. Five years of frustration followed, and Samuel finally decided to run a blind ad in the June 1, 1954, *Wall Street Journal.* It read:

> The Four Dead Sea Scrolls: Biblical Manuscripts dating back to at least 200 B.C. are for sale. This would be an ideal gift to an educational or religious institution by an individual or group. Box F 206, The Wall Street Journal.

Either by coincidence or divine providence, E. L. Sukenik's son, Yigael Yadin, an archaeologist and Israeli political figure in his own right, happened to be in America at the time. One day, he received a call from a friend, who had noticed this unusual ad in the *Wall Street*

Journal. Could it be that these were the very scrolls his father had hoped to buy years earlier, from the Metropolitan's emissary? Convinced that this was the case, he proceeded to negotiate the purchase of the scrolls, through a bank acting as a "neutral" broker. In this way, Yadin managed to acquire, for a sum of $250,000, all four scrolls, on behalf of the State of Israel. A total of seven Dead Sea Scrolls were now in possession of the Jewish state. So valued were these parchments that a special wing of the Israel Museum of Jerusalem was constructed – the Shrine of the Book. The museum directors clearly understood that people need positive national images to which to rally, and the fledgling nation of Israel was fortunate to have many of them: the Temple, the Tabernacle, the Ark of the Covenant, the menorah, and now the Dead Sea Scrolls.

Yigael Yadin, Chief of General Staff

For quite some time, it was thought that there were no more parchments to be found in the desert of Judea, but then came the discovery by the Bedouin of additional manuscript finds in the lonely cliffside caves. In short order, there began a virtual race to scour each nook and cranny along the shores of the Dead Sea, at Khirbet Qumran. It amounted to what might be called the "Great Judean Desert Scroll Rush." The only real question was: who will get there first, the archaeologists, or the Bedouin? In 1949 the archaeologists made their own concerted effort, in an expedition organized by Father Roland de Vaux and Gerald Harding of the Jordanian Department of Antiquities. There were also expeditions from the prestigious *Ecole Biblique*, the Palestine Archaeological Museum, and the American School of Oriental Research. By then, however, this entire area had fallen to the Jordanians, who promptly annexed it. Neither Israelis nor any Jews were invited to take part, and none of the finds were made accessible to Jewish scholars. Archaeology had become captive to a new, anti-Israel drumbeat.

In any case, the new discoveries easily bordered on the spectacular. The caves yielding scrolls or fragments were numbered according to the chronology of their discovery. In Qumran Cave 3 a most curious artifact was retrieved. It was unlike all the others, which have been written on dried animal skins. This item consisted of a roll of relatively pure, highly refined copper, broken in two, and so oxidized that the archaeologists were unable to open it. Having no other recourse available, they had it sawn into a series of arced segments, whereupon its mysterious contents were painstakingly deciphered. They consisted of a lengthy inventory of buried treasure, perhaps the treasure of the great temple in Jerusalem, along with clues describing the precise locations. A subsequent expedition to locate the treasure, again with no Jews involved, turned up not a trace, and to this day, the very nature of the Copper Scroll remains a mystery.

Dead Sea Copper Scroll

Another discovery was uncovered in what was dubbed Qumran

Cave 11. It contained some of the best preserved scrolls, including the longest of them all. It amounts to a detailed description of an enormous temple, to stand on Jerusalem's Mount Moriah (Temple Mount) at the end of time. It is appropriately called The Temple Scroll. Yet another cave, perhaps the most important of all, became known as Qumran Cave 4. Discovered by the Bedouin, it was home to tens of thousands of fragments from over five hundred different documents. Some of the most important of these include:

- a letter, apparently written by the founder of the ancient sectarians who lived in the region,
- fragments of another rule book for the sect, called the Damascus Document,
- assorted ancient prayers and blessings, and
- previously unknown, unread commentaries on the biblical books of Isaiah, Psalms, Hosea, Nahum, and Habakkuk.

Unfortunately, the Bedouin had hardly been trained in the techniques of archaeology, and, more often than not, they simply grabbed large numbers of delicate parchment fragments, with no thought for their fragile condition. In archaeology it is vitally important that artifacts be found *in situ*, that is, "in place," without being moved or removed. This is essential in establishing their so-called "provenance," or precise place of origin. The Bedouin, however, took no note of where, within a particular cave, the fragments had been found, frustrating to no end subsequent attempts to piece them together. Some manuscript fragments were even broken into smaller pieces, in the desire to have more items to sell. All in all, serious damage was done in those early years to the scrolls being retrieved. Ultimately, the Kingdom of Jordan decided that the best way to cajole the Bedouin to turn in their scraps of parchment was to buy them, notwithstanding that they had pilfered the parchments illegally. The official price was the equivalent of $2.50 per square inch of inscribed surface. Not surprisingly, it worked. Over time, more than forty thousand fragments were turned in, from Cave 4 alone.

In its next order of business, the Palestine Archaeological Museum (now called the Rockefeller Museum) assembled, in September of 1952, a committee of eight scholars, which was given the daunting task of piecing together and deciphering the mountain of fragments. Academics of great renown were selected, including Frank Cross, of Harvard Divinity School. Of course, no Israeli scholars were represented, nor was there a single Jewish scholar on the team. Was this political, anti-Semitic, or both? Riding herd over the committee was Harvard's John Strugnel, who, in spite of his reputation as an eminent scholar, subsequently gave an interview to an Israeli newspaper in which he betrayed certain anti-Semitic sentiments. He eventually stepped down from the committee, citing health reasons.

In any case, the rush to obtain more scrolls was fierce, making it all the more difficult for scholars to examine them once assembled. In addition to the caves in the vicinity of young Muhammad's original find, near the ruin of Qumran, the expeditions of the coming years uncovered scroll material at such locations as the Wadi Ed-Daliyeh, Wadi Murabba'at, Nahal Hever, Nahal Se'elim, and the great fortress-rock that rises from the Judean desert, Masada. Moreover, since the scrolls were certainly written by an assortment of scribes, and since some predated the Qumran settlement itself, new ideas arose regarding who the authors might have been.

Qumran Caves

Were they the so-called Essene sect, who embraced an apocalyptic mentality and secluded themselves from the rest of society? Were they a radical group of "Sadducee" priests? Were they a proto-Christian group, who may have represented an alternate form of early Christianity, or some other group entirely? According to other theories, the scrolls may have been smuggled out of Jerusalem prior to it being besieged by the Romans during the Great Revolt of 66 to 70 C.E. In that case, might the scrolls represent at least a metaphysical connection between the ancient Jewish drive for independence and the modern quest for a Jewish state? That of course would be of great interest to Israeli Jews, but there would be no Jews involved in this research, at least for the time being.

Digging Up a Mound

During the early 1950s the riddle of the Dead Sea Scrolls took a new turn, when Father Roland de Vaux, under the supervision of the Jordanian government, organized a major excavation of the slumbering ruins at Khirbet Qumran, which lay in close proximity to the caves themselves. De Vaux was a meticulous scholar as well as a Roman Catholic priest. He was a veteran in the region, part of the Catholic Church's permanent presence in the land of Jesus. It took him six grueling years to excavate the ruins, and what he found was nothing less than astounding. As the diggers meticulously removed the rubble from the site, an entire ancient settlement began to appear. Unsurprisingly, no Jews were involved in the excavation. Moreover, most of de Vaux's finds would never be published. Nonetheless, a general "picture" of the site began to emerge.

It appears that the earliest settlement at what came to be called Qumran was founded in the eighth through seventh centuries B.C.E., toward the end of the First Temple period, thanks to traces of a modest fort or fortified farmhouse found in the ruins. It may possibly have been the place known as Secacah, or the City of Salt, referred to in the book of Joshua:

> In the wilderness: Beth-arabah, Middin, and Secacah; and Nibshan, and the City of Salt, and En-gedi; six cities with their villages. (Joshua 15:61-62)

That early structure was restored and enlarged in the late second century B.C.E., most likely during the rule of Hasmonean King John Hyrcanus. Later, under King Alexander Yannai, in the early first century B.C.E., the plan of the site evolved, with additional settlements and construction. That included an aqueduct leading from a cliff above Wadi Qumran, a few hundred meters away. Floodwaters during the winter rains filled a reservoir behind a dam at the foot of the cliff, and fed the many cisterns, including what were identified as ritual immersion baths, or *mikva'ot*.

The layout of the site is uniquely dissimilar to other settlements of that period, featuring multiple large halls, likely designed for public functions, and a limited number of living quarters. The settlement's main entrance was at the foot of a watchtower, to the north. The buildings' walls consisted of stones gathered at the foot of the cliff, and were covered with thick, white-gray plaster. The doorposts and windows were made of well-trimmed stones, the roofs being composed of wooden beams, straw and plaster, as was common at that time. There were multiple cisterns and ritual baths – *mikva'ot* – found in various locations, covered with gray hydraulic plaster for waterproofing. They were supplied with water from the aqueduct. There was a wide descending staircase with a plaster-covered ridge in the middle, separating those going down for immersion from those coming up purified. One would imagine that Jewish and Israeli scholars would be interested in such findings, but politics would prevail, and for Jews politics has rarely been friendly.

Qumran

De Vaux's excavations also determined that in the year 31 B.C.E., an earthquake seriously damaged the buildings, including the cisterns and *mikva'ot*. Cracks in the walls and a thick layer of ash from a fire were the telltale signs of catastrophic damage. Afterwards, the site lay abandoned, until the early first century C.E., at which time members of the community returned and resettled it. The earlier structures, with various additions and modifications, were re-inhabited. However, in the year 68 C.E., during the Jewish revolt against Rome, the settlement was completely destroyed, never again to be occupied.

With regard to who the inhabitants were, and to which group they belonged, mention should be made of de Vaux's "smoking gun." In the main building was a long room with a number of plaster tables and benches, which gave every indication of being furnishings used by what he conceived as the ancient Jewish "monastic" order known as the Essenes, who, he was convinced, had inscribed the scrolls. De Vaux came upon the tables in a long room which he called the "scriptorium." In his mind it was here that monk-like scribes labored day and night, while others constantly read aloud the sacred texts in shifts. De Vaux was certain that he had come upon the "headquarters" of the Essene sect. It was later questioned whether de Vaux, himself a Dominican monk, may have been drawn to possible parallels between these mysterious texts and Christianity. When he conceived of this site as an ancient "monastery" of sorts, was he perhaps stamping his own preconceptions on it?

Further muddling his conclusions, it should be noted that a large cemetery of over one thousand graves was discovered, arranged in north-south rows and extending across the marl surface adjacent to the settlement. Several were excavated, revealing simple graves dug one by one into the marl, each covered with a pile of stones. Most of the skeletons were male, though toward the edge of the cemetery, graves of females and children were also uncovered. Does this sound like a "monastic" settlement? Indeed, there is nothing more foreign to Judaism as a religion than the concept of celibacy. Might Jewish archaeologists, had any been involved in the dig, have come to different conclusions? In the geopolitical climate of the 1950s, however, that was never an option. The one thing perfectly clear was that everything about the settlement related to Jewish life and culture in antiquity. How odd, that its excavation was in the hands of a Catholic cleric, to the exclusion of Jews.

Nonetheless, after all the years that passed since de Vaux's excavation of the 1950s, the proverbial jury is still out on the ruins of Qumran. Every conclusion has been and remains debatable. In 1988 two Belgian archaeologists returned to Qumran, to reexamine the findings of de Vaux. Even before publishing their findings, they gave an interview, on the Public Television program *Nova*, devoted to the scrolls, charging that the entire site was not an ancient "monastery" at all, but a wealthy Roman villa. As for the plaster writing tables de Vaux had found, they identified them as dining room tables. They further argued that several elegant ceramic urns found by de Vaux actually belonged to a perfume and cosmetic industry located in the region during Roman times. They claimed that the Jewish immersion baths were only cisterns, and that the communal dining room was actually just an assembly hall, common to villas of the period. Another theory was put forth by Professor Norman Golb of the University of Chicago. He concluded that Qumran was neither a monastery nor a villa, but a military fortress, ultimately put to use by the first century "Zealot" sect, fighting for freedom from Roman rule.

Under New Management

The Dead Sea Scrolls "status quo" was destined to change drastically, however, with Israel's conquest of Judea and Samaria in the 1967 Six-Day War. At last Jewish and Israeli scholars were given full access to the site, as well as the thousands of scroll fragments sequestered in the Rockefeller Museum. It literally took a war to bring a modicum of equilibrium to the archaeology of Qumran. More recently, two Israeli archaeologists, Itzhak Magen and Yuval Peleg, never shy of stirring the scholarly "pot," theorized that the settlement was in actuality an ancient pottery factory. Qumran, they insisted, had nothing to do with the Dead Sea Scrolls found in the nearby caves. Magen declared:

> Your vision of a couple hundred celibate Essenes padding around praying whenever they were not copying scrolls in a special room designated "the scrollery," only to end up buried in a silent cemetery of more than a thousand single graves, is a work of the imagination, not history or archaeology...

Qumran Scriptorium/ Scrollery

What was Qumran? According to Magen, it was, in its first phase, a Hasmonean fortress built to protect the eastern frontier of the kingdom. Later it was simply a pottery factory, manned by a few dozen workers at most. This conclusion, he insisted is "inescapable."[7] Magen did not dispute the contention that at least two of the cisterns excavated by De Vaux fit the description of ritual immersion baths (*mikvaʾot*), but he noted that the largest of these cisterns holds in excess of three hundred cubic meters of water. That is enormous, compared with other *mikvaʾot* found in other parts of ancient Israel. He also pointed out that just before the points where the water would enter a number of the cisterns, there are hollows, where sediments would likely have collected. That would have made the cisterns unkosher for use as *mikvaʾot*. Magen argued that they must have been used for something else entirely.

What about the hundreds of clay dishes found in the room identified by De Vaux as the pantry? Magen argued that they have not been completely accounted for. We have to ask why so many were found intact, and why there were so many to begin with, when far fewer would have sufficed to feed everybody at the settlement. Magen and Peleg contend that these dishes were in fact the chief product of the site. Furthermore, in the process of digging out the debris that filled the large cistern, they discovered a fine layer of clay, amounting to roughly three tons. This clay is what was fashioned into dishes and fired in the two large ovens found in the excavations. That would explain the substantial number of dishes uncovered – in a room that was not a pantry, but a storage room for finished merchandise. It would also rule out Qumran as a place of learning and contemplation, since it would have been impossible to study in the midst of a loud, dirty factory. This, they contend, was the site's *raison d'être*: a fortress, which, after the Roman conquest in 63 B.C.E. and the dissolution of the Hasmonean army, became, at the hands of the out-of-work soldiers, a pottery factory.

Such arguments are of course hotly contested by leading scholars, but at least they come from a different perspective – a Jewish one. It is pointed out that the pottery found at Qumran is in fact of a poor and common grade, which would not be expected in a wealthy villa, much less in a pottery production center. Furthermore, while there is no doubt that pottery was produced at Qumran, there is also no evidence that it was "marketed" anywhere else.

As for the contention that Qumran was a fort, it should be noted

---

7   Hershel Shanks, "Qumran – The Pottery Factory," *BAR* 32/5 (2006).

that the settlement has no fortress-like wall, and the outer walls are no thicker than the inner walls. Finally, there are the scrolls themselves. We really cannot dismiss the sheer proximity of the caves to the ruins at the site of Qumran. After Israelis finally gained access to the entire area, curator emeritus of the Shrine of the Book, Magen Broshi, noted that Cave 4, with its vast horde of parchments, is situated just across the ravine from the settlement itself. This, he argued, simply cannot be co-incidental. There must be a connection; the scrolls must have been com-posed there, or at least gathered there. Broshi took the additional step of excavating the ancient trails leading back and forth from the caves to the settlement, uncovering abundant remains, such as fragments from an-cient sandals and even teeth from two thousand year-old combs. Clear-ly, there was a good deal of human traffic linking the site of Qumran with the caves, and that is a serious problem for those who claim that whoever lived at the settlement had nothing to do with the scrolls.

Certainly, there is no question as to the Jewishness of the inhabitants of Qumran. Pottery shards have been discovered at the site, inscribed with Jewish names, as well as stone vessels used by Jews of the period in order to abide by the Jewish purity laws. As scroll research opened up under Israeli authority, many others have joined the debate, from all backgrounds, including Jewish scholars. Prof. Yizhar Hirschfeld, of the Hebrew University of Jerusalem, agreed that the earliest Second Temple phase of Qumran, around 100 B.C.E., amounted to a large fortified Has-monean tower. But Hirschfeld argued that the cisterns were drinking water pools, and that the inhabitants who used them were farm work-ers, possibly producing highly valued balsam essence, or cultivating date palms between the settlement and the spring of Ein Feshka to the south.[8] According to Hirschfeld, Qumran was an estate belonging to Judea's elite class, most likely members of the priesthood, and the scrolls were the remains of a large Jerusalem library – possibly the temple li-brary. They were smuggled from Jerusalem to Qumran, whose inhab-itants were sympathizers or even family. They were then deposited in caves, from which they could be retrieved at the end of the Great Revolt.

For his part, Itzhak Magen detailed how he believed the scrolls came to be deposited in the Qumran caves:

---

8 By contrast, Magen pointed out that these cisterns were dug inside the building complex, after the initial phase of the compound had already been constructed. Why would cisterns be placed inside the complex when the inhabitants could just as easily have dug them outside? The only reason would be if the settlement had been industri-al, and the cisterns part of the production process.

They were brought here by everybody, including fugitives running away from the Romans. Some of them would have taken a scroll with them, but when they ran away from the Judean hills eastwards, they had to cross the water, which is something they didn't want to do with a scroll.

Magen suggested that it was these fugitives who sequestered the scrolls among the caves in the vicinity of the now-deserted settlement of Qumran. That would mean that they were not priestly or sectarian compositions at all. As Magen pointed out:

This is the literature of Second Temple era Judaism. This belonged to everybody.[9]

The amazing reality is that, even after two long millennia, the Dead Sea Scrolls and the legacy they convey still belong to everybody, and that especially includes the Jewish people. Hopefully, with the tools of archaeology at our disposal, everyone may know a great deal more about who lived in the barren wilderness known as Khirbet Qumran.

In the meantime, not a few modern Israelis have transcendental moments as they reverently file into Jerusalem's Shrine of the Book Museum. There, unrolled, around the inner rim of an enormous circular glass case is the complete scroll of the book of the Prophet Isaiah, which speaks to them across the millennia – of dispersion, suffering, inquisition, exile and Holocaust. As their little Israeli children sound out the words of the prophets of old in Hebrew, their mother tongue, they know in an instant that they are intimately connected with the Jews of that distant epoch. Not a few modern Israelis are brought to tears in the presence of these ancient texts, that resonate as mute witnesses, saying, "We Jews are still here!"

"The Scrolls are Coming!"

Occasionally, an assortment of Dead Sea Scrolls and fragments is sent on tour to various countries, showcasing the archaeological richness of the State of Israel. This is excellent PR as well as a lucrative source of revenue. It would be no exaggeration to say that the scrolls are the equivalent of archaeological "rock stars." An analogy might be to Elvis Presley's gold Cadillac, which was sent on tour to Australia and

---

9 Yaron Ben-Ami, "The Enigma of Qumran," *The Bible and Interpretation* (Oct. 2004): https://bibleinterp.arizona.edu/articles/Ben-Ami--The_Enigma_of_Qumran.

New Zealand in the late 1960s. With newspaper headlines announcing, "The Car is Coming!" it was reported that there had never been lines so long. Similarly, an exposition of the Dead Sea Scrolls, which opened in Toronto in 2009, was met with public acclaim, incredible interest and long lines. Unlike the case of Elvis's car, however, all of the publicity was not positive. Protesters showed up on a daily basis, waving flags of the Palestinian National Authority. Their boisterous chants betrayed their agenda: the Royal Ontario Museum was displaying stolen property![10] Not surprisingly, they had a legal case to make. The general area where the Dead Sea Scrolls were found, Qumran, was at the time part of the Emirate of Transjordan, which was absorbed into the British Mandate of Palestine. When the first scrolls were discovered in 1947, the area was still under British administration.

Dead Sea Scrolls in the Rockefeller Museum

Jordan formally annexed this territory in 1949. The Department of Antiquities of Jordan, in tandem with East Jerusalem's *École Biblique*, organized collaborative expeditions from 1947 to 1956, and the thousands of fragments retrieved by the Bedouin from Cave Four were gathered by Jordanian officials in the East Jerusalem Rockefeller Museum.

---

10 See Robert R. Cargil, "On the Curious Protests of the Dead Sea Scrolls Exhibition in Toronto," *Bible and Interpretation* (Aug. 2009): http://www.bibleinterp.com/opeds/curious.shtml.

With Israel's victory in the 1967 Six-Day War, however, East Jerusalem was annexed to the Jewish state, and with it the Rockefeller Museum and its contents, including its treasure trove of scrolls. The pro-Palestinian demonstrators at the Toronto scroll exhibition accused Israel of illegally moving the parchments from the Rockefeller Museum to the West Jerusalem Shrine of the Book Museum. The charge is absurd on its face, since the major scrolls on display in the Shrine of the Book are the ones legally purchased by Sukenik and Yadin. The mountain of scroll fragments remain in the Rockefeller, even though it is now under new, Israeli "management."

Wars of course have consequences, but even if it were assumed that Israel had no right to the Rockefeller Museum, to whom should its contents now be repatriated? The Palestinian Authority did not exist at the time the scrolls were discovered. If any national entity other than Israel has a claim on these parchments, it would be Jordan, which ruled the area from 1949 to 1967. Who, however, gave Jordan the right to annex Judea and Samaria, including East Jerusalem and Qumran? Moreover, the Dead Sea Copper Scroll, with its inventory of buried treasure, resides today in a museum in Amman, Jordan. Where is the charge that it has been "looted?" Where is the demand from the Palestinian Authority that it too must be repatriated? Though the claim is often made that Israel uses archaeology for political purposes, it is clear in this case that the pro-Palestinian side is using the archaeology of Qumran for its own political purposes. Elvis's car was never given such a hard time.

In the final analysis, we should ask to whom are the Dead Sea Scrolls important and why they are valuable. They were written many centuries before the prophet Muhammad was born, long before Islam existed. The faith of Muhammad has multiple monuments and many millions of adherents, worldwide. The Jews have only Israel, a sliver of a country no larger than the state of New Jersey. The Jewish people are not remembered for their splendid cities of antiquity. Their settlements contained no marble-faced amphitheaters, such as those of the Romans, who cheered to the bloodletting of gladiatorial combat. The Jewish legacy, by stark contrast, was portable. It consisted of holy books, fashioned not from exquisite marble, but leathery vellum. That is what the Dead Sea Scrolls represent, and that is why they belong, all politics aside, to the people of Israel.

The outline of the ancient guard tower at Qumran today stands sentinel against the deep cerulean sky. But what is it guarding? The ancient Jews who lived there are long gone, but their written legacy thunders across the millennia. It is the past, as preserved in their sacred texts, which must be guarded, since, as we well know, the past is prologue. The ancient Jewish scribes, who labored to produce this priceless library, could hardly agree more.

EPILOGUE –
PROTECTORS OF HISTORY

I N 2010 THE ARAB television network Al-Jazeera produced a documentary conveying a viciously anti-Israel, propagandistic screed.[1] The Jewish state is accustomed to being excoriated in innumerable media "hit pieces," but this one was unique, in the sense that it blamed Israelis for (as the title of the piece proclaimed) "Looting the Holy Land." Israelis are said to be actively pilfering antiquities from sites in the "West Bank"/ "Palestine," and delivering them to private collectors, who make enormous profits from their sale. It even recycles the old charge that the tunnels Israel has opened in Jerusalem's Old City are undermining the al-Aqsa Mosque.

The film is also careful to highlight self-critical Israelis, such as archaeologist Jonathan Mizrahi, a former employee of the Israel Antiquities Authority. A vocal critic of the City of David excavations and the archaeological park built to showcase them, he observes, "You come in here, and you hardly see any Palestinians in your tour." But is this not an ancient Jewish site by nature? Is the absence of Palestinians not a simple reflection of the fact that there were no Palestinians here that long ago? That reality is of course never mentioned. Palestinian archaeologist and academic, Mahmoud Hawari, observes, "Palestinians in general see archaeology and the archaeological sites as part and parcel of Israeli occupation." However, if this is how Palestinians choose to see archaeology, is that not their own problem? The program's narrator then adds, "The Palestinians have not used archaeology to shape their national identity." Shall we laugh now or later?

In response, archaeologist Israel Finkelstein offered a passionate re-

---

1   Al-Jazeera, "Looting the Holy Land," https://www.aljazeera.com/programm es/2010/10/201010591416403822.html.

buttal:

> I wish to demonstrate – point by point – why this is a worthless film,
> ridden with manipulations, political propaganda, incorrect facts and
> even lies.[2]

He points out that while the Palestinian director of archaeology is
featured in the documentary, not a single representative of the Israel Antiquities Authority is given a voice. Such bias, he notes, is all too typical
in the Arab world. Israelis are depicted as either ultra-religious settlers,
soldiers, or gun-brandishing zealots, while the Palestinians come off
as "… peace-loving farmers riding donkeys in beautiful fields with ro-
mantic flute music playing in the background."[3] Furthermore, observes
Finkelstein, fully half of the "West Bank" has been under Palestinian
control since the 1993 Oslo Accords, and whatever looting has occurred
has been carried out under the gaze of the Palestinian Authority.

Looters, Libel and Lies

Indeed, most of the archaeological looting has been perpetrated, not
by Israelis, but by Palestinian Arabs, who have perpetrated their banditry in Israel proper as well as Judea and Samaria. The documentary
goes as far as to libel Israel's first Prime Minister, David Ben Gurion,
whose interest in archaeology is linked with the so-called "expulsion" of
Arabs from their residences during the 1948 War of Independence. He
is accused of wanting to excavate on their land, even though his interest
in archaeology only developed in the 1950s, well after the Arabs became
refugees. Finkelstein asks rhetorically:

> Do people expect that the Jews will focus on the good old days in their
> respective one hundred eighty countries but ignore the importance of
> their historical homeland, many parts of which lie underground and
> require excavation to find and scientific study to understand?[4]

In short, the al-Jazeera "hit piece" is one more example of how Israel is accused of politicizing archaeology, while the political agenda of
the Palestinians escapes any serious scrutiny. Summing up, Finkelstein
comments:

---

2 Finkelstein, "'Looting the Holy Land' or Pillaging the Truth?," op. cit.
3 Ibid.
4 Ibid.

Archaeology in Israel is prospering and is of a high quality, ranking among the best in the world. Israeli archaeologists are in the front-line of research. A hasty look at the front pages of international journals which deal with archaeology – not only of the Levant – reveals a number of Israeli scholars far beyond their proportion in the world archaeology community. With all criticism that one (once in a while by this author, too) may have concerning the Israel Antiquities Authority, the organization administers archaeology in Israel in an efficient, orderly and professional way. And finally, without counting articles, I dare state that Israeli scholars contribute to the knowledge of Islamic archaeology more than all archaeologists in the Arab world combined. In the end, archaeology – as every other science – is decided by the level of education and scholarly work, not by politics and propaganda.[5]

In short, Israeli archaeologists, far from being "looters" of history's priceless artifacts, are in truth their protectors. When it comes to propaganda, another conclusion is easily drawn. Namely, its mode of dissemination at the hands of the Palestinian Authority is so pandemic that the bulk of the population of Judea, Samaria and Gaza honestly does not know that once upon a time this entire area was Jewish land. It is as if the Jewish people have been entirely written out of history. Yet, in antiquity, Israel truly was a "Greater Israel," far larger than the sliver of a country that hugs the Mediterranean coast today. Not only in the so-called "West Bank," but in today's Hashemite Kingdom of Jordan, there are innumerable "facts underground" – subtle reminders that all of this land was once thoroughly Jewish.

Nevertheless, the Jordanian government is clearly "weaponizing" archaeology against their Jewish neighbors, and writing Jews out of history altogether. The name "Israel" is not referenced on any archaeological maps or markers, notwithstanding that they are subsidized by the American organization USAID. As for Jews, multiple archaeological sites in the Hashemite Kingdom advance the political agenda that two millennia ago, they were already the "oppressors" of eastern Mediterranean lands. An example relates to the Jewish dynasty of the first two centuries B.C.E., the Hasmoneans, who threw off the yoke of their Syrian ("Seleucid") overlords, and reigned as an independent Jewish nation until conquered by the Roman legions of Pompey in 63 B.C.E. A sign in the Jordan Archaeological Museum in Amman explains:

> After the Seleucids achieved domination over the entire area from the

---

5   Ibid.

late 3rd Century BC onwards, the militant Hasmonean Jews rose up against Greek domination and established their own reign in Palestine and the Northern part of Jordan. Most of the Greek cities welcomed the roman army headed by General Pompey as a liberator from Jewish oppression.[6]

At the spectacular archaeological site of Petra, where temples are carved into cliffs, a sign in the newly renovated museum describes the Hasmonean King Alexander Yannai, or "Janneus:"

> Alexander Janneus was King of Judah and he was a ruthless ruler who sought to expand and strengthen the territories of Judah. Around 100 B.C. he took control of Gaza and though the people of Gaza asked for Aretas [of Petra] help it came too late.[7]

There are many other promising locations east of the Jordan River where archaeologists might find remains of ancient Jewish presence. From the days of David and Solomon, the "Levitical cities" in Transjordan included:

- Ramoth-gilead
- Mahanaim
- Jazer
- Mephaath
- Heshbon
- Bezer
- Jahzah, and
- Kedemoth.

Even the capital of modern Jordan, Amman, was once Israelite, known as Rabbath-bene-ammon. Yet, it is hardly likely that such sites, or a host of others, will be opened to Israeli archaeologists any time soon. There is an Israeli nursery rhyme that goes: "Two banks has the river Jordan; one is ours and the other also." The "facts underground" indeed support the reality that all of this territory once belonged to the people of Israel. When it comes to the international community, however, the modern Jewish state is not even to be given leave to govern the "West Bank" of the Jordan River, even though this territory was won in a defensive war for its very survival – even though the alternative would be the creation

---

6 Adam Sacks, "How Jordan Is Weaponizing Archaeology Against Israel," *Haaretz,* Apr. 1, 2019: https://www.haaretz.com/israel-news/.premium-vengeful-expansion-ist-false-jordan-finds-ancient-roots-for-hostility-to-israel-1.7048617.

7 Ibid.

of a radical terrorist state, bent on pushing its Jewish neighbors into the sea.

Since 1967, Israel has at least been able to explore Judea and Samaria, as well as the Golan Heights, where a significant number of major ancient sites are scattered. For Jews, such places amount to more than archaeological excavations; they are the very reflection of the "Jewish soul," frozen, as it were, in time. How much of the "Jewish soul" will be lost if or when the "heart" of ancient Israel becomes a Palestinian state – off limits to the very Israelis who for decades have excavated these places and the many Jews who have sought to live in them again? Do we really imagine that they will be respected and cared for by an independent and possibly Hamas-dominated "Palestine?"

A Personal Note

Some years ago I attended the summer institute of Yad Vashem, the national Holocaust museum of the State of Israel, in Jerusalem. A distinguished professor from the Hebrew University of Jerusalem provided a compelling analogy. Suppose you leave your house for an extended vacation. When you return, you find that the door is locked. After ringing the doorbell, someone else answers. You say,

"This is my home!"

The intruder says,

"No it isn't! This is my home; I live here and I have always lived here."

You say,

"I'm sorry, but you are mistaken. I built this house and it is sitting on my own land."

The intruder says,

"You are wrong. I own this house and the land on which it was built. You must leave immediately!"

At this point what recourse do you have? Perhaps you will go and fetch the blueprints to establish that this is the house you designed. Is it reasonable, given your predicament, for someone to object, that you are bastardizing architecture for your own selfish interests? So it is with archaeology. The purpose is not to cheapen it or to bastardize it or politicize it. However, when the leadership of the Palestinian Arabs, relative "Johnny-come-latelys" on the landscape of historical Israel, make serious claims that this land is not Jewish at all, then it might reasonably be argued that archaeology is not being "used;" it is being put to legitimate use to establish whose "house" this really was, and is.

When I first came to Israel, as a young undergraduate student of history from the University of Illinois at Chicago, I received my first intro-duction to biblical archaeology, while studying at an institute for Amer-icans on Jerusalem's Mount Zion. On a weekly basis, my classmates and I were escorted to multiple archaeological sites across the length and breadth of Israel and its territories.

There was free and open access to archaeological sites, even in the most tense areas. I was able to visit Jericho, Hebron, Shechem, and scores of other locations. It was easy to spot Israeli troops on patrol, or perched on rooftops, watching over the surrounding territory. Why were they present? To "oppress" the Arab population? No. They were there for one reason, to prevent the outbreak of terrorism. It was clear to me at the time that if the Arab population was not intent on promoting terror, if they chose to cooperate with their neighbors and build a prosperous, multicultural society, no security precautions would be necessary: no checkpoints, no roadblocks, no security fence, no "profiling" of poten-tial threats.

Since then, the creation of the Palestinian Authority hardly helped. Locations I used to visit are places I would no longer go. I am glad that I had the chance to visit so much archaeology in Judea and Samaria, as I have no idea when or if I will ever see such places again. When I visit Is-rael today, I can easily spot the separation barrier, and I know at a glance what it means. Israelis cannot live in peace with neighbors who want to murder them. I sincerely hope that the barrier will someday be disman-tled; but it will come down, not with the creation of a hostile, Palestinian state, but with the extension of Israeli law and sovereignty (along with full Israeli citizenship for the Arabs) over the whole land of Israel, not just a truncated sliver. As the father of modern Zionism, Viennese-born Theodore Herzl, once asserted, "If you will it, it is not a dream!"

Final Thoughts

Archaeology, like any science, feigns neutrality, even if no such thing exists among mere mortals, who always have a proverbial horse in the race. Observable data are invariably used to draw inferences and conclusions, hopelessly tinged with the color of the investigator's biases. The astronomer peers through a telescope to theorize how the universe began; the excavator turns over earth with a spade to determine who lived when and where. Archaeologists, like cosmologists, must of ne-cessity be consumed with the reconstruction of the past, whether by examining the wispy spirals of ever-receding galaxies or by studying the faint outlines of ancient walls on the desert floor.

Climbing down the cliffs of time, both astronomer and excavator are tasked with producing plausible interpretations based on immutable science. In each case the objects of the quest, be they the whispering voices of ancient civilizations inscribed in stone or the pulses of background radiation emanating from the Big Bang, are the stuff of intellectual speculation mingled with a degree of "artistic license." The cosmologist imagines the process of creation itself; the archaeologist conjures up the hanging gardens of lore and fragrances of honey and myrrh filling the air. Neutrality, in the final analysis, is an illusion. The important thing is not to be neutral but to be right. Hopefully, these pages have, at least to some small degree, advanced the Israeli cause, the Jewish cause in the Middle East – not because Israel is perfect, but because it is "right."

Index

236

239

244

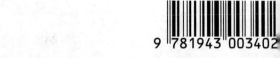